BACTERIOLOGY RESEARCH DEVELOPMENTS

BACTERIOPHAGES

BIOLOGY, APPLICATIONS AND ROLE IN HEALTH AND DISEASE

Bacteriology Research Developments

Additional books in this series can be found on Nova's website
under the Series tab.

Additional e-books in this series can be found on Nova's website
under the e-book tab.

Microbiology Research Advances

Additional books in this series can be found on Nova's website
under the Series tab.

Additional e-books in this series can be found on Nova's website
under the e-book tab.

BACTERIOLOGY RESEARCH DEVELOPMENTS

BACTERIOPHAGES

BIOLOGY, APPLICATIONS AND ROLE IN HEALTH AND DISEASE

CLARK DENTON
AND
RICHARD J. CROSBY
EDITORS

NOVA BIOMEDICAL

New York

Library of Congress Cataloging-in-Publication Data

Bacteriophages: Biology, Applications and Role in Health and Disease / Editors, Clark Denton and Richard J. Crosby.

pages cm

Includes index.

ISBN: 978-1-62618-513-5 (hardcover)

1. Bacteriophages. 2. Bacteriophages--Therapeutic use. 3. Genetic vectors. I. Denton, Clark editor of compilation. II. Crosby, Richard J editor of compilation.

QR342.B3382 2013

579.2'6--dc23

2013009433

Published by Nova Science Publishers, Inc. † New York

Contents

Preface

Bacteriophages are obligate parasites that require bacterial hosts to replicate. Phages initiate infections by first attaching to the surface of bacteria and injecting their genomes into host cells. Virulent phages then hijack the host's cellular machinery to direct the production of phage virons that burst from the cell leading to bacterial lysis. In this book, the authors discuss the biology, applications and role in health and disease of bacteriophages. Topics include the varied potential roles of bacteriophages in the production of food from animal sources; phage therapy, biocontrol and commercial microbiology; abortive infection as a killer bacteriophage resistance strategy; bacteriophages as biocontrol tools of pathogens in meat and dairy products; bacteriophages as probiotics and decontaminating agents for food products; use of bacteriophages as surrogate indicators of viruses in water; promoter sequences for bacteriophage RNA polymerases; the use of bacteriophages in bacterial food pathogen regulation; and the interaction between bacteriophage and its bacterial host receptor, LamB.

Chapter 1 - Antibiotic resistance in pathogenic bacteria has heightened interest in alternative antimicrobial agents for the control and treatment of bacterial infections. Bacteriophages (phages) are one such potential biological control agent. Bacterial infections, both clinical and subclinical, have an enormous impact on the health and productivity of production animals, thereby limiting the profitability of these enterprises. The presence of certain bacterial strains also has implications for food safety and public health. Reliance on antibiotics for treatment and control of bacteria is likely to become less feasible over time, not just due to resistance but also an increasing public demand for organic food and products free from preservatives and other chemicals. Phage therapy may be one alternative to lessen the requirement for antibiotics in animal production. Bacteriophages have co-evolved with bacteria, so they are already present in most environments, including those involved in animal production. Ruminant animals, such as cattle and sheep, even carry bacteriophages in the rumen, which is not unexpected given the high bacterial load within this biological fermentation system. There has been considerable research into the use of bacteriophages for control of pathogenic bacteria in the major terrestrial production animal species – cattle, sheep, pigs and poultry. Even more has been directed at production in aquaculture, covering aquatic organisms such as fresh-water fish, crustaceans and molluscs. In mollusc and shrimp aquaculture, for example, bacterial infections can have devastating consequences, but the use of antibiotics can be problematic, so bacteriophages represent a potential alternative control mechanism. There are already bacteriophage products on the market that are licensed for use in products destined for human consumption. These bacteriophage products focus on food safety, however, rather than *in vivo* treatment of bacterial infections. These may have commercial benefit in terms of the public perception of a natural, preservative-free product. The presence of bacteriophages in food production, however, is not always favourable. In the dairy industry, for example, bacteriophages of lactic acid bacteria are of enormous negative economic concern. Infection of milk samples with bacteriophages results in lysis of the starter culture cells and interferes with the milk fermentation process. So, in dairy production, the priority is to exclude or control them in the milk. It is clear that bacteriophages already have a biological role in aspects of animal production, both naturally and through research development. Research indicates that they may have an expanded role in animal production, both in terms of the health and productivity of the animals and in producing food that is safe for human consumption.

Chapter 2 - The first studies involving phage dating from the early nineteenth century. In the last three decades these studies were intensified, mainly because of the emergence of antibiotic-resistant bacterial strains. Since then, the wide and diversity of phage species became evident. It is estimated that there are 10^{31} particles and 10^{8} different species of phages dispersed in the biosphere, which highlights their importance in ecosystems on the planet. These data generates the need for an efficient phage classification system. Classically, the phages have been classified according to their structure, type of genetic material (DNA or RNA), host range and infection cycle. Additionally, molecular techniques such as pattern of enzymatic restriction of the DNA, profile of structural proteins, and more recently gene sequencing have been increasingly employed for this purpose. The original studies involving phage aimed the control of pathogenic bacteria in humans and animals, named the phagetherapy. Based on phage estimated number and diversity, it is likely that there are thousands of phages with potential to be applied in several areas. The promising results encouraged the researcher to search for alternative applications such as food control, biofilm dispersion in capillaries and surfaces, specific detection of microorganisms and even as instruments of gene transference for bacterial cells. Thus, a proportional amount of information has been and will still be generated. The gathering of these data in an organized way will support the creation of a viable classification system for efficient implementation of phagotherapy in the prevention and control of pathogens that threaten human and animal health.

Chapter 3 - Bacteriophages are the most abundant biological entities on earth and outnumber their bacterial hosts by a factor of 10. As a result, bacteria have evolved numerous defense systems to resist viral infection. These systems operate at all stages of the phage life cycle and include cell surface modification, restriction-modification and CRISPR-mediated resistance. Another notable defense strategy is phage abortive infection (Abi), where a bacterial host prevents the production of viable phage progeny and in the process protects the bacterial population as a whole, but at the expense of its life. For this reason, phage Abi can be viewed as "altruistic suicide". Recently, a link has been demonstrated between abortive infection activity and the near-ubiquitous prokaryotic toxin-antitoxin (TA) systems. Here, the authors review abortive infection and the emerging role for TA systems in mediating this powerful anti-viral activity.

Chapter 4 - Biocontrol by bacteriophages in meat and dairy products was developed in the last years and their recent "generally recognized as safe" (GRAS) designation and approval as food additives (FDA, 2006) has opened the discussion about "edible virus". The phage treatment can be applied at different stages in the food chain production, from the live animal to the manufactured product. In this chapter the authors focus on their application during milk and raw meat processing prior to consumption in order to extend the shelf life of these perishable foods. Studies using bacteriophages as biocontrol tools in meat products have focused particularly on *Escherichia coli* O157:H7, but other pathogens such as *Campylobacter jejuni*, *Listeria monocytogenes* and *Salmonella enterica* have also received attention. Even though the major concern in dairy production is preventing phage infection of dairy starter cultures, the aim of this chapter is to address the use of bacteriophages for biocontrol of pathogens like *Staphylococcus aureus*, *Escherichia coli*, *Enterobacter sakazakii*, *Listeria monocytogenes*, *Mycobacterium smegmatis* and *Salmonella enteritidis*. On the other hand, a section will be dedicated to revise the biocontrol of Shiga-toxigenic *Escherichia coli* (STEC) pathogens in meat and dairy foods. STEC is the etiologic agent of hemolytic uremic syndrome (HUS), a disease endemic in countries like Argentine, which has the highest incidence worldwide with 13.9 cases per 100,000 children under 5 years of age. In spite of the advantages of the use of bacteriophages in food safety, a drawback is the emergence of bacteriophage insensitive mutants (BIMs). Nevertheless, low frequency of emergence, phenotype reversions and variable stability of BIMs were reported to date. Furthermore, phage cocktails can be used to counteract the effect of BIMs occurrence.

Chapter 5 - Bacteriophages are components, determinable quantitatively, of natural microflora of food production continuum which covers all production stages: from growing agricultural plants and animals to retail sales of ready food products. The viruses of bacteria are stably present in all objects of the environment and may be easily extracted from soil specimens, fresh and sea water, washouts from plants surfaces, human and animals excrements, ready for sale food products. Sterile and cleansed of bacterial toxins the phage lysates with high titers of viral particles are used for species-specific decontamination of food products in all stages of manufacturing process and storage and are employed as probiotic dietary supplement that allow to reduce the risk of food-borne infections caused by *Escherichia*, *Salmonella*, *Shigella* and other pathogenic bacteria, including their antibiotic resistant strains. Protection of food products from bacterial contamination using the strategy of phage biocontrol is a highly differentiated preserving technique which has prolonged effect, is safe for humans and is relatively inexpensive as compared to commonly used techniques.

Chapter 6 - There are many potential applications of bacteriophages as models to assess the behaviour and survival of enteric viruses in the environment, and as surrogates to assess the resistance of human viruses to water treatment and disinfection processes. To minimize pathogens health risk, members of the coliform bacterial group, including *Escherichia coli*, in drinking water are measured to determine whether water has been contaminated by fecal material. Nevertheless, viruses are more resistant to chlorine and other disinfectants used in water treatment than are coliforms. Thus, fecal coliform numbers do not serve as adequate indicators of enteric virus contamination which may affect human health after consumption of products contaminated by reclaimed water or sludge. The use of bacteriophages as indicators of the presence and behaviour of enteric bacteria and animal viruses has always been attractive because of the ease of detection and low cost associated with phage assay and for similarity to enteric viruses in terms of transport and survival characteristics. Somatic coliphages, F-specific RNA phages, and bacteriophages that infect *Bacteroides fragilis* have been proposed as potential indicators of water quality and as tracers for enteric viruses presence in water due to their viral nature and their similar size and structure.

Chapter 7 - Because of their high transcription efficiency and easiness of manipulation, bacteriophage RNA polymerases are commonly used for *in vitro* transcription. In addition, their applicability for *in vivo* expression has also been demonstrated. Thus, bacteriophage RNA polymerases are promising components for artificial *in vitro* and prokaryotic expression systems. To widen the variety of independently regulated transcription machineries, the generation of bacteriophage RNA polymerase mutants with altered promoter specificities has been explored. Strict orthogonality of these mutants is required for practical applications, and therefore, the promoter specificities of the mutant RNA polymerases should be validated. For the analyses of promoter sequences, *in vivo* systems have been traditionally employed, but recently, several *in vitro* systems, by which huge numbers of promoter sequences can be analyzed with fewer complications, have been also developed. In this chapter, I review the progress in the development of experimental systems for the analyses and engineering of promoter sequences for bacteriophage RNA polymerases.

Chapter 8 - Contemporary technologies that are employed to inactivate bacterial pathogens in foods are not flawless; thus, novel approaches for improving food safety are needed. Moreover, consumers desire more natural foods, i.e., free from chemical preservatives. Bacteriophages, which are viruses that infect and kill bacteria, have garnered recent interest for use in food safety. Bacteriophages are components of the natural microflora that are present throughout food production, from the farm to the retail outlet. They are stable in these environments and are readily recovered from soil, sewage, water, farm and processing plant effluents, feces, and retail foods. Studies that have assessed the use of bacteriophages to control food pathogens have yielded promising results but require further improvement. The success of this methodology is ultimately based on consumer acceptance; thus, careful consideration of phage origin must be performed before introducing bacteriophages into foods.

Chapter 9 - With the recent emergence of antibiotic resistant bacteria, there is an urgent need for new therapeutic alternatives. Bacteriophages are bacterial viruses that provide an attractive alternative to treat bacterial infections. One of the key factors for a successful phage therapy is the high affinity binding of phages onto the bacterial cell surface. This very first step of virus-host interaction would determine subsequent infection steps and the fate of the entire infection process. Phage λ and its host *Escherichia coli* serve as an excellent paradigm to study the mechanism of phage adsorption onto the host cell surface receptor. Here the authors review their current understanding of the mechanism of lambda phage adsorption with an emphasis on recent imaging studies using fluorescently labeled phage λ and its host receptor LamB.

In: Bacteriophages
Editors: Clark Denton and Richard J. Crosby
ISBN: 978-1-62618-513-5

Chapter 1

Bacteriophages from Farm to Fork: The Varied and Potential Roles of Bacteriophages in the Production of Food from Animal Sources

***J. Owens*[*,1] *and K. R. Hodgson*[2]**
[1]Faculty of Veterinary Science, The University of Melbourne, Parkville, Victoria, Australia
[2]Food Safety and Innovation, Plant Research Centre, Urrbrae, Australia

Abstract

Antibiotic resistance in pathogenic bacteria has heightened interest in alternative antimicrobial agents for the control and treatment of bacterial infections. Bacteriophages (phages) are one such potential biological control agent. Bacterial infections, both clinical and subclinical, have an enormous impact on the health and productivity of production animals, thereby limiting the profitability of these enterprises. The presence of certain bacterial strains also has implications for food safety and public health. Reliance on antibiotics for treatment and control of bacteria is likely to become less feasible over time, not just due to resistance but also an increasing public demand for organic food and products free from preservatives and other chemicals. Phage therapy may be one alternative to lessen the requirement for antibiotics in animal production. Bacteriophages have co-evolved with bacteria, so they are already present in most environments, including those involved in animal production. Ruminant animals, such as cattle and sheep, even carry bacteriophages in the rumen, which is not unexpected given the high bacterial load within this biological fermentation system. There has been considerable research into the use of bacteriophages for control of pathogenic bacteria in the major terrestrial production animal species – cattle, sheep, pigs and poultry. Even more has been directed at production in aquaculture, covering aquatic organisms such as fresh-water fish, crustaceans and molluscs. In mollusc and shrimp aquaculture, for example,

* Corresponding author: Telephone: +61 3 9035 9706; Fax: +61 3 8344 7374; Email: jane.owens@unimelb.edu.au.

bacterial infections can have devastating consequences, but the use of antibiotics can be problematic, so bacteriophages represent a potential alternative control mechanism. There are already bacteriophage products on the market that are licensed for use in products destined for human consumption. These bacteriophage products focus on food safety, however, rather than *in vivo* treatment of bacterial infections. These may have commercial benefit in terms of the public perception of a natural, preservative-free product. The presence of bacteriophages in food production, however, is not always favourable. In the dairy industry, for example, bacteriophages of lactic acid bacteria are of enormous negative economic concern. Infection of milk samples with bacteriophages results in lysis of the starter culture cells and interferes with the milk fermentation process. So, in dairy production, the priority is to exclude or control them in the milk. It is clear that bacteriophages already have a biological role in aspects of animal production, both naturally and through research development. Research indicates that they may have an expanded role in animal production, both in terms of the health and productivity of the animals and in producing food that is safe for human consumption.

Introduction

Animal agriculture accounts for a significant percentage of world agricultural production. In the United States alone livestock and poultry contribute to over half of US agricultural cash receipts, in excess of $100 billion per year [1]. There are 90.8 million cattle and 5.35 million sheep in the United States [2, 3]. In Australia there are 25.9 million cattle and 73.1 million sheep [4], more than four times the size of the total human population in the country. With the human population expanding globally, and an estimated 10 billion mouths to feed by the end of the century [5], there is an ever increasing need for efficient and sustainable agriculture. Protein from animal sources is one of the important components of the diet for most populations. Clinical and subclinical diseases, often caused by bacterial infections, are limiting factors to the productivity of animal agricultural enterprises. With growing global concern in regards to the use of antibiotics and spread of antibiotic resistance, the administration of antibiotics to food animals will be more tightly governed. The emergence of resistant strains of *Campylobacter* in poultry eventually led to the US Food and Drug Administration banning the use of fluoroquinolones in poultry in 2005 [6]. In the United States, the European Union, Australia and other parts of the world antibiotic use in animals is heavily regulated [7, 8].

Pathogens that cause clinical illness or limit animal productivity are not the only concern in animal production. Zoonotic pathogens are of enormous importance in terms of safe food production. Many outbreaks of foodborne disease can be traced back to an animal source [9, 10]. One pathogen of concern is *Escherichia coli*. *E. coli* O157:H7, and other strains, have been responsible for numerous outbreaks worldwide [11]. This organism can be traced back to animal sources. It is often found in the gastrointestinal tract of cattle and other animals and therefore is a risk for contamination of food products, both directly through the contamination of the carcass but also indirectly such as through contamination of water that is used to irrigate food crops [12, 13]. Strategies to minimize the presence of this and other bacterial pathogens in the animals themselves are highly sought after in the effort to produce safe food products. Alternatively, effective products to treat food products further along the chain of production have potential, such as meat spray treatments for deli meats, which can be contaminated by pathogens such as *Listeria* [14-17].

Bacteriophages (phages) are one potential biological control agent for bacterial pathogens. They are naturally occurring entities that are found in the presence of their host bacteria. It is their host-specific ability to infect and lyse bacteria that make them an attractive target for research into the control of bacterial pathogens [18]. The economic importance of animal agriculture worldwide and the impact of bacteria both in terms of direct animal illness, loss of production and human health impacts mean that research into the use of bacteriophages in animals and animal products is escalating. It is desirable, given the issues of antibiotic resistance in human and animal bacteria, that alternatives are available for control of these pathogens. There is also an increasing public demand for organic food and products free from preservatives and other chemicals [19]. Bacteriophages are found in the majority of environments where bacteria are found, so they are naturally occurring and may meet the desire for bacterial control without the addition of chemicals. They have co-evolved with bacteria (both pathogenic and non-pathogenic) and are naturally present in animal production environments. Bacteriophages can be found in the gastrointestinal tracts of different animal species. Ruminants, such as cattle and sheep carry bacteriophages in the rumen, which is not unexpected given the high bacterial load that acts as a biological fermentation system for the digestion of plant material [20, 21].

In recent times there has been an increasing volume of research into the potential use of bacteriophages in animals. Terrestrial animal species such as cattle, sheep, pigs and poultry, representing a large proportion of animal agricultural production, have been the focus of much of this research. Cattle and poultry are the main species to which the research has been directed. Some research has looked at controlling animal pathogens which cause clinical disease and or loss of production in the animals directly, but there has also been considerable focus on the potential use of bacteriophages in the food safety area. It is not just the terrestrial species to which bacteriophage research has been directed. There has been extensive research in the area of aquaculture. Bacterial infections can have devastating impacts in aquaculture production enterprises, but the use of antibiotics can be problematic often owing to the use and requirement to dispose of large quantities of water [22]. Studies have been carried out in aquatic organisms such as fish, crustaceans and molluscs.

The impact of bacteriophages can be deleterious. They exist in a symbiotic relationship with bacteria suggesting the bacteria are tolerant to their presence, therefore considerable research is required to harness their properties adequately. Temperate bacteriophages have the ability to incorporate into the genome of their host bacteria and are capable of transferring genes between bacteria [23]. This may be beneficial in terms of the use of bacteriophages to transfer lethal genes to bacteria, but it also means they are capable of transferring other undesirable genes such as those that encode for toxins or resistance [23-25]. The presence of bacteriophages is not always desirable. In the dairy industry bacteriophages of lactic acid bacteria can interfere with the milk fermentation process [26]. Therefore, exclusion of these bacteriophages is very important in dairy production. Bacteriophages already have a natural biological role in animal production and considerable research with promising results can be presented to suggest that they may have an expanded role in the future. There is still much research to be done to ensure the safety and efficacy of bacteriophage therapies for the treatment or control of animal pathogens or in terms of food safety.

Ruminants

Ruminants, including beef and dairy cattle, sheep, goats, buffalo and deer, account for the majority of world animal agricultural production [27]. The United States is the world's largest beef producing country [28] and, including buffalo and goats' milk, India is the largest milk producing country with approximately 94 million tonnes produced annually [29]. The extraordinarily large numbers of livestock and the level of production mean that control of disease is imperative, but exclusion of certain bacteriophages from dairy production is also important. The host-specific nature of the bacteriophages, however, means that phage control of pathogens should not impact on the starter cultures. The bacterial pathogens of concern in ruminant production are genetically very different to the lactic acid bacteria important in the fermentation process of dairy production.

Beef Cattle

One of the most significant pathogens investigated extensively in beef cattle in terms of potential bacteriophage control is *E. coli*. Both in terms of the health of the animals themselves, primarily in calves, and human health, *E. coli* pose a significant risk. Cattle are reservoirs for numerous different strains of *E. coli*, including the Shiga toxin producing *E. coli* (STEC) O157:H7 [30, 31], which has been responsible for many foodborne disease outbreaks in humans causing serious clinical illness and death [32, 33]. Niu et al. studied bacteriophages of *E. coli* O157:H7 in cattle feedlots. Of 855 samples tested they were able to isolated bacteriophages active against *E. coli* O157:H7 in 239 of the samples. They observed that there was some reduction in faecal shedding of *E. coli* O157:H7 in pens where bacteriophages could be isolated from the cattle in the same pen [34]. Niu et al. also investigated the host range and lytic capabilities of four bacteriophages against *E. coli* O157:H7 isolates, which included 297 bovine isolates. Based on *in vitro* experiments, the bacteriophages were shown to have varying host ranges and were capable of lysing up to 96.4% of strains tested. The susceptibility of the bovine strains was shown to be influenced by the genotype, as determined by pulsed-field gel electrophoresis, and the phage typing phenotype [35].

Rozema et al. compared the efficacy of oral and rectal administration of bacteriophages active against *E. coli* O157:H7 for reducing faecal shedding of the pathogen in steers. Four bacteriophages were administered as a cocktail either orally or rectally, and in combination. There was a numerical reduction in the number of nalidixic acid-resistant *E. coli* O157:H7 cultures from the orally treated group, but there was no difference in the total number of *E. coli* O157 strains obtained from the control and treatment groups. The control group also shed bacteriophages, despite not being treated with them, which the authors suggest may be due to natural acquisition from the environment [36]. Investigation of oral delivery of bacteriophages was also reported by Stanford et al. They commented on the need to protect orally-administered bacteriophages against gastric acidity and reported on the use of polymer encapsulation to deliver bacteriophages via the oral route. Four bacteriophages were polymer-encapsulated (termed "Ephage") and delivered either top-dressed on feed or in gelatin capsules to cattle that had been artificially infected with *E. coli* O157:H7. Ephage did not reduce the levels of faecal shedding of *E. coli* O157:H7, but there was a 14-day reduction in the duration of shedding over a 10-week period [37].

Another study isolated bacteriophages active against enterohaemorrhagic *E. coli* (EHEC) and STEC strains. The bacteriophages were isolated from minced beef and bovine faecal samples. This study focussed on a number of important considerations in the use of bacteriophages *in vivo*. Bacteriophages that were capable of lysing non-pathogenic *E. coli* were excluded from the experiment as they are considered to have potentially detrimental effects on the normal gut microflora. Selection of bacteriophages included screening for undesirable genes including the Shiga toxin encoding genes (stx1 and stx2) and the c*I* gene [38], which encodes a protein that interferes with the lytic cycle and thereby favours the lysogenic cycle [39].

Studies have also investigated the use of bacteriophages to control *E. coli* O157:H7 on cattle hides to mitigate the risk of transfer of the organism to humans. Coffey et al. showed that two bacteriophages (e11/2 and e4/1c) could be effective in reducing the numbers of *E. coli* bacteria on contaminated hide samples. As with the study by Dini and de Urraza [38], Rivas et al. considered the importance of detecting any undesirable genes such as those that may encode for virulence or toxin genes, and the genomes of the bacteriophages used were sequenced completely [40]. The same bacteriophages were also tested in an *ex vivo* rumen model and in cattle *in vivo*. Bacteriophage e11/2 was capable of reducing the numbers of *E. coli* O157:H7 in the artificial rumen system to below detectable levels within 1 hour. Similar results were not able to be achieved, however, in the *in vivo* experiments. There was no difference in *E. coli* O157:H7 shedding by the treated and untreated animals [41]. These results demonstrated the complications of transferring bacteriophages that are effective in experimental models into living animals.

In earlier work, O'Flynn et al. investigated the efficacy of a bacteriophage cocktail against O157:H7 on beef carcasses. The bacteriophages were shown to be effective in significantly decreasing the level of contamination [42]. The use of bacteriophages for control of *E. coli* on the hides of livestock has developed commercially. OmniLytics™ (Salt Lake City, Utah) has developed a bacteriophage product for the treatment of cattle hides, on live cattle, prior to processing to control *E. coli* O157:H7. The United States Food and Drug Administration (FDA) issued a no objection letter for the use of the product in 2007. The product is also designed for treatment of transportation vehicles and cattle holding areas. It is applied as a mist, spray or wash [43]. The same company has also developed a range of products branded as BacWash™ for similar applications to control *Salmonella*. The FDA also issued a letter of no objection for the use of these products in 2007 [44]. These were the first bacteriophage products to be approved for use in live animals [45].

Aside from the human health implications of certain bacterial pathogens in animals, many also pose a health risk to the animals themselves. Diarrhoea is a common cause of disease and often death in young beef and dairy calves [46]. Enterotoxigenic *E. coli* (ETEC) is a common cause and use of bacteriophages as a potential control measure has been investigated. Calves were treated orally with a bacteriophage cocktail and a reduction was seen in the levels of *E. coli* in faecal samples from the calves in the bacteriophage treatment group compared to the treatment group, although the difference was not statistically significant [47].

Dairy Cattle

There are numerous reports on the potential use of bacteriophages to control pathogens associated with dairy cattle. Mastitis is the most significant disease of dairy cattle and can present as either clinical or subclinical disease [48]. One of the common causes of mastitis in dairy cows is *Staphylococcus aureus* [49, 50]. One study reports the isolation of anti-staphylococcal bacteriophages from farmyard slurry. These bacteriophages were shown to be able to lyse *S. aureus* isolates from clinical cases of mastitis *in vitro*. They also showed weak lytic activity against some human strains of methicillin-resistant *S. aureus* (MRSA). The study also demonstrated that administration of the bacteriophage cocktail into the udder did not result in an increase in the somatic cell count and there was no evidence of a local immune response [51]. Another study isolated a bacteriophage from the milk of a cow with clinical mastitis. This bacteriophage showed lytic activity against *S. aureus* isolates of bovine and human origin in *in vitro* trials [52]. Bacteriophage endolysin has also been tested *in vitro* against *Staphylococcus* isolates from cows with mastitis [53]. Gill et al. also investigated the use of a lytic *S. aureus* bacteriophage. Cows with subclinical *S. aureus* mastitis were treated with intramammary infusion of the bacteriophage. There was no difference in the cure rate between the control and phage-treated groups. An increase in the somatic cell count was observed in healthy lactating cows administered the bacteriophage via the intramammary route, but not in cows with existing *S. aureus* infection that were also administered the bacteriophage [54].

Other significant conditions that are important in dairy cows are post partum diseases – metritis and endometritis [55]. One study tested the sensitivity of *E. coli* isolated from the uteri of postpartum dairy cows to environmental bacteriophages, as a possible treatment for post partum uterine infection. The bacteriophages showed efficacy in lysing *E. coli* strains *in vitro*, including multi drug resistant strains of *E. coli* [56, 57]. Another study also reports the intrauterine administration of a bacteriophage cocktail to dairy cows. The uterine health and fertility of the dairy cows was monitored, with a particular focus on the isolation of *E. coli* and *Arcanobacterium pyogenes* from the uteri. *In vivo* the bacteriophage treatment did not have any measurable effect on the results of uterine secretion cultures or reproductive performance, and it did not prevent metritis or endometritis [58].

Dairy cattle are also a potential source of zoonotic pathogens, many of which have been discussed for beef cattle. One such bacterial pathogen is *Listeria monocytogenes*, which is common in dairy farm environments [59]. Bacteriophages active for *L. monocytogenes* – "listeriaphages" – have been isolated from silage samples on dairy farms. This particular study reported that listeraphages were more abundant in the silage than *L. monocytogenes* isolates. The sensitivity of *L. monocytogenes* to the listeriaphages was reportedly high, with individual bacteriophages differing in their ability to lyse the stains with some having a narrow and others a broad host range [60]. Control of *L. monocytogenes* on dairy farms is beneficial to human health due to the foodborne disease risk of the organism [59].

Dairy Produce

In the manufacturing process of milk and other dairy products, the use of bacteriophages has been investigated for a number of applications. Bacteriophages have been shown to reduce *Salmonella* Enteritidis development during the manufacture and storage of Cheddar cheese [61]. In red smear cheeses, bacteriophages have been applied for the control of *L. monocytogenes* growth [62], and in curd manufacturing processes bacteriophages have been applied to control the growth of *S. aureus* [63]. Another study also investigated the use of bacteriophages to control *S. aureus* in milk. They demonstrated bacterial growth inhibition in ultra heat treated (UHT) and pasteurized whole-fat milk, but less activity in semi-skimmed raw milk and very little activity in whole raw milk. Bacterial lysis was observed at room temperature and at 37°C, but was not successful at refrigeration temperature [64]. The same study reported a high prevalence of *S. aureus* bacteriophages in dairy samples (bulk tank milk and cheeses) from dairy cows. All the bacteriophages isolated were, however, temperate bacteriophages which are not desirable for prophylactic or treatment applications [64].

The presence of bacteriophages in dairy production is not always positive. Bacteriophages of lactic acid bacteria can interfere with the fermentation process in the manufacture of dairy products and these bacteriophages are capable of surviving for long periods of time in production facilities [65]. Lactic acid bacteria are essential in the fermentation process of dairy product manufacture. These bacteria ferment lactose resulting in the production of lactic acid, which decreases the pH. The manufacture of cheese and fermented milk products are reliant upon this process and it is also essential in controlling the growth of other pathogenic and spoilage bacteria [66]. The bacteriophages of dairy bacteria have been quite extensively reviewed [65, 67, 68]. Fortunately, the necessary exclusion of lactic acid bacteriophages from dairy manufacturing does not necessarily limit the use of bacteriophages in other dairy-related applications, as most bacteriophages are species and even strain specific [18].

Sheep

Research into bacteriophage use in sheep is not as extensive as cattle. A number of papers report the investigation of the bacteriophage population in the rumen environment. Iverson and Millis demonstrated variability over a period of 112 days in the sensitivity of rumen *Streptococcus bovis* bacteria in fistulated sheep to bacteriophages, in the absence of detectable bacteriophages [69]. Rumen strains of bacteria have naturally variable sensitivity to bacteriophages. The bacteriophage population in the rumen has been shown to be highly labile and unique to individual animals, even those cohabitating where the potential for transfer of rumen bacteria and bacteriophages theoretically exists [70]. Cheong and Brooker isolated a virulent bacteriophage specific for *Propionibacterium* spp. from the rumen fluid of sheep. They also demonstrated that the diet of the sheep influenced the ability to isolate bacteriophages as they were isolated from sheep on a protein-rich lucerne diet but not from sheep fed on a high-fibre, low-protein diet [71]. The results of the dietary influence on the bacteriophage population are supported by other work [72].

Other research has demonstrated a lysogenic bacteriophage carrying the Shiga toxin (stx_{1OX3}) gene, which is found in some *E. coli* isolates of ovine origin. Healthy sheep can be carriers of STEC isolates, which is of concern to humans as a foodborne pathogen and is frequently a cause of the deadly haemolytic uraemic syndrome [73]. This highlights the concern of lysogenic bacteriophages and their potential to transfer genes associated with toxin production. Changes in the bacterial genome of *E. coli* in sheep via bacteriophage transfer are considered to be a significant risk [74]. Other *E. coli* strains are thought to be able to acquire *stx*-carrying bacteriophages, thereby making them a more significant risk to human health [75]. It is not considered, however, that bacteriophage carrying the Shiga toxin encoding genes has any influence on the ability of the strains to preferentially colonize sheep [76].

Control of *E. coli*, specifically O157:H7, has been investigated by other researchers. Use of a bacteriophage in an artificial rumen system (Rusitec) had some inhibitory effect on the growth of *E. coli* O157:H7, however a similar result was not observed in lambs [77]. One study showed a decrease in faecal shedding of *E. coli* O157:H7 in the presence of a bacteriophage (CEV1), which was replicated experimentally with reduced levels of *E. coli* O157:H7 in the caecum and rectum following treatment with CEV1. The bacteriophage was active against 17 out of 19 strains of *E. coli* O157:H7 that were tested in the study [78]. Bacteriophages have a significant role in the transfer of genes in organisms of zoonotic risk to humans, but also a potential role in the control of the same organisms.

Pigs

In comparison to a large proportion of sheep and cattle production, pigs are often raised intensively. This creates more challenges for the control of bacterial pathogens as transmission is often facilitated by the close cohabitation of the animals. Pigs are also carriers of a number of significant zoonotic bacterial pathogens. *Salmonella* is one such pathogen. The presence of *Salmonella* on farms and at a pig slaughterhouse has been reported. Using both *S.* Typhimurium and *S.* Heidelberg as isolation strains, bacteriophages were isolated from effluent manure both on-farm and at the slaughterhouse at a very high rate. *Salmonella* strains isolated at the same localities were found to be resistant to many of the isolated bacteriophages [79]. *Salmonella* isolation has shown a marked increase in pigs just prior to slaughter, especially with increased transport and lairage times, which is a risk for foodborne transmission [80, 81]. Wall et al. demonstrated a 2-3 log reduction in *Salmonella* in the tonsils, ileum and caecum of young pigs concurrently inoculated with *S.* Typhimurium and a bacteriophage cocktail (both via oral gavage). The bacteriophage cocktail consisted of 15 bacteriophages grown individually on a *S.* Typhimurium strain, purified and then encapsulated as a cocktail. The reduction seen in market-weight pigs was not as marked, but still showed a statistically significant decrease in *Salmonella* in the caecal contents [82]. One research group has also demonstrated a decrease in *Salmonella* shedding post-challenge in experimental animals direct-fed a microencapsulated bacteriophage cocktail, as well as in those pigs administered the same bacteriophage cocktail via oral gavage [83]. A bacteriophage cocktail has also been tested on the skin of pigs to reduce contamination with *S.* Typhimurium U288. Application of the bacteriophage cocktail to pig skin artificially contamination with the U288 strain showed a reduction in the numbers of *Salmonella* present on the skin, stored at 4°C, over 96 hours [84].

E. coli is also a significant pathogen of pigs, capable of causing significant clinical disease in the pigs themselves and posing a zoonotic risk. One study demonstrated a reduction in the overall duration of diarrhoea and mean diarrhoea scores in pigs experimentally infected with *E. coli* O149 and either administered bacteriophages prophylactically or as a treatment (24 hours post challenge). A significant reduction was not, however, observed with all the bacteriophages administered individually [85]. According to the analysis method using diarrhoea score [86] another study also showed a reduction in enterotoxigenic *E. coli* in pigs treated prophylactically with either 10^6 plaque forming units (PFU)/g or 10^8 PFU/g of a bacteriophage prior to oral challenge with *E. coli*. The bacteriophage in the treated pigs was mixed with feed one week prior to the *E. coli* challenge [87].

The potential use of bacteriophages as a vaccination delivery system has also been investigated. Recombinant bacteriophages have been used to deliver antigens for vaccination against *Taenia solium* pig cysticercosis. Three peptides (KETc1, KETc12 and GK1) and a recombinant antigen KETc7 were expressed on the surface of the M13 bacteriophage. Preliminary investigation showed a potential protective effect from the vaccine [88]. Oral vaccination against the highly contagious foot-and-mouth disease virus (FMDV) using a capsid protomer vaccine displayed on T4 bacteriophage surface has also been tested. Capsid precusor polyprotein (P1) and proteinase 3C from a pig strain of the virus were expressed on the T4 phage surface. Oral delivery of the bacteriophage resulted in up to 100% protection in experimental infection in mice and two vaccinated pigs did not develop infection after 6 months contact with susceptible pigs, in which FMDV was observed [89].

Other microorganisms have also been investigated. Shi et al. report the use of two bacteriophage lytic proteins against *Streptococcus suis*. In combination the two proteins showed lytic activity against 15 of 19 *S. suis* isolates tested [90]. Zehr et al. investigated a bacteriophage of virulent *Haemophilus parasuis*. This Mu-like bacteriophage is thought to potentially to contribute to the transfer of virulence genes between strains [91]. Yan et al. also investigated the use of bacteriophage supplementation as potential growth promoters, as an alternative or adjuvant to antibiotics. No significant effects were observed, except for a reduction in the concentration of *E. coli* isolated from the pigs [92].

Poultry

Bacteriophage research in poultry is quite extensive. Numerous targets have been identified for bacteriophage control, including bacterial pathogens of poultry and those posing a risk of foodborne infection to humans. One pathogen of concern to humans is *Campylobacter*. Poultry are the primary reservoir for *Campylobacter* spp., especially *C. jejuni*, and are the primary source of cases of human infection [93]. Bacteriophages have been isolated from raw commercial chicken meat, indicating that they are able to survive processing and storage and may be suitable control agents for microorganisms in poultry that pose a foodborne disease risk [94].

One study demonstrated the efficacy of a bacteriophage cocktail to target both *C. jejuni* and *C. coli* in live birds. The bacteriophage cocktail was administered via oral gavage and a reduction in *Campylobacter* levels in the birds was observed. Similar results were observed when the bacteriophage was given in the feed, although the reduction in *Campylobacter* levels was observed to occur sooner than for the oral gavage treatment. Thirteen percent of *Campylobacter* strains isolated were seen to be bacteriophage-resistant [95]. El-Shibiny et al. also demonstrated a reduction in colonization of chickens with both *C. jejuni* and *C. coli* using a single bacteriophage (CP220). The rate of identification of bacteriophage resistant strains in this study was lower, however, with only 2% of strains isolated determined to be bacteriophage resistant following exposure to the virulent bacteriophage [96]. Scott et al. demonstrated that *C. jejuni* strains that survive bacteriophage predation can undergo genomic rearrangements that can contribute to bacteriophage resistance [97]. The author of this review also reports the isolation and characterization of *C. jejuni* bacteriophage from indoor and free range poultry. Bacteriophages from free range poultry were shown to be effective in lysing a range of *C. jejuni* strains *in vivo* [98]. Research into *Campylobacter* bacteriophages and their use in poultry has been thoroughly reviewed [99].

E. coli is a commensal organism in the gastrointestinal tract of poultry, but some strains of extra-intestinal avian pathogenic *E. coli* (APEC) are capable of causing serious infection outside the intestinal tract. The disease is referred to as colibacillosis [100, 101]. Numerous experiments have investigated the use of bacteriophages to control *E. coli* in poultry. Oliveira et al. report the isolation and characterization of bacteriophages from poultry sewage that are capable of lysing APEC strains [102]. They then tested the *in vivo* efficacy of a bacteriophage cocktail to control severe respiratory infection due to *E. coli* in both artificially and naturally infected chickens. The bacteriophage cocktail was administered as a single application, orally and via a spray. A dose-dependent response was observed in decreasing morbidity and mortality in experimentally infected birds, but the application of the bacteriophages at a lower titre was successful in decreasing morbidity and mortality in naturally infected birds [103]. Jamalludeen et al. also report the isolation and characterization of bacteriophages active against *E. coli* serotypes O1, O2 and O78, which are the main causative agents of colibacillosis in poultry [104].

Barrow et al. tested the efficacy of a lytic bacteriophage in decreasing morbidity and mortality in *E. coli* septicaemia and meningitis-like infection in chickens. Mortality was close to 100% in untreated chickens. When birds inoculated with *E. coli* were concurrently administered the bacteriophage via intramuscular injection (in different muscles) no illness or deaths were observed. The result was dose-dependent. Intracranial inoculation with *E. coli* and concurrent bacteriophage treatment also reduced morbidity and mortality in the bacteriophage-treated group. With intracranial inoculation with *E. coli* the results of treatment with the bacteriophage were also dose-dependent. Concurrent administration of the infective agent and bacteriophage does not, however, represent the likely timeline of infection with a pathogen, leading to clinical signs that can result in diagnosis/identification of illness, followed by treatment. Significantly, however, the bacteriophages in this study were shown to have multiplied in the blood [105]. The bacteriophage treatment in the experiments by Lau et al. was delayed by two hours. In these experiments the birds were inoculated with *E. coli* and the bacteriophage treatment administered both via intratracheal inoculation. The bacteriophage-treated birds had lower levels of *E. coli* and mortality in the treated group was significantly reduced [106].

One study reports a decrease in colibacillosis mortality from 68% to 15% with bacteriophage administration via the intramuscular route [107]. The same research group later reported that prior exposure to a bacteriophage is likely to limit the ability of the same bacteriophage to be a successful treatment option. Chickens that had been pre-treated with the bacteriophage prior to exposure to *E. coli* and subsequent bacteriophage treatment were found to have higher anti-bacteriophage IgG levels than birds that were not pre-treated with the bacteriophage. Serum from these birds inhibited bacteriophage activity [108].

Salmonella is another significant bacterial genus associated with poultry. The Salmonella genus contains two species – S. enteric and S. bongori – and six subspecies – I enterica, II salamae, IIIa arizonae, IIIb diarizonae, IV houtenae, and VI indica [109]. Certain *Salmonella* serovars are associated with disease in poultry and also pose a zoonotic risk via foodborne transmission. Fiorentin et al. isolated bacteriophages from free range chickens that were lytic for *Salmonella* enteric serovar Enteritidis (*S.* Enteritidis). This research demonstrated a reduction, by 5 days post-treatment, in *S.* Enteritidis in the caecal contents of artificially infected birds that were treated therapeutically with three different bacteriophages [110, 111]. *S.* Enteritidis poses a significant foodborne disease risk from poultry products [112]. Another study isolated a large number of *Salmonella* bacteriophages from poultry farms, abattoirs and waste water. Three of these bacteriophages had broad host ranges against *Salmonella enterica* serotypes Enteritidis, Hadar, and Typhimurium. The bacteriophages were successful in reducing the levels of colonization of experimentally colonized chickens with both *S.* Enteritidis and *S.* Typhimurium. No reduction in *S.* Hadar was observed. It was observed that at higher titres the bacteriophages induced a higher rate of isolation of bacteriophage-resistant *Salmonella* [113]. Hurley et al. were not able to demonstrate a reduction in the shedding of *Salmonella* with oral administration of a single bacteriophage, however this study used mathematical modelling to examine the potential efficacy and impediments to bacteriophage therapy [114]. Eggs are another foodborne source of human *Salmonella* infection and survival of bacteriophages in artificially inoculated eggs has also been demonstrated [115], suggesting that bacteriophages may have the potential to control bacterial pathogens in eggs.

Salmonella enterica serovar Pullorum (*S.* Pullorum) can cause severe disease in poultry, with high morbidity and mortality. It is a common disease of poultry in Africa and Asia [116, 117]. Bao et al. isolated and characterized two bacteriophages specific for *S.* Pullorum, which showed significant lytic ability against different strains *in vitro* [118]. *Salmonella enterica* serovar Gallinarum (*S.* Gallinarum) is another significant pathogen of poultry, causing fowl typhoid which is an acute septicaemic disease [119, 120]. A significant reduction in mortality was observed in chickens artificially infected with *S.* Gallinarum that were treated with bacteriophage 7 days before and 21 days after *S.* Gallinarum inoculation [121].

Bacteriophage control of other bacterial pathogens in poultry has also been studied. One such example is a study by Miller et al. that investigated the potential use of bacteriophages to target *Clostridium perfringens* [122]. *C. perfringens* causes necrotic enteritis in poultry, which can result in high morbidity and mortality [123]. A bacteriophage-derived, toxoid-type vaccine and a bacteriophage cocktail were separately assessed for their efficacy in controlling disease due to *C. perfringens* in artificially-infected broiler chickens. Chickens treated orally with the bacteriophage cocktail had significantly lower mortality due to necrotic enteritis than untreated, infected birds. The bacteriophage cocktail was reported to be more effective than the toxoid vaccine. A significant reduction in mortality and morbidity was observed by all methods of oral bacteriophage administration tested – oral gavage, via feed or drinking water [122].

Aquaculture

Aquaculture is a rapidly expanding industry and control of pathogens presents significant challenges. Use of prophylactic and therapeutic antibiotics has increased and antibiotic resistant strains of bacteria have emerged. The problems associated with the use of antibiotics in aquaculture have been reviewed by Cabello [124]. Bacteriophages are considered a viable adjunct to antibiotic use in aquaculture. Nakai and Park reviewed the use of bacteriophages in cultured fish [125]. More recently, the use of bacteriophage therapy as a control strategy in the broad aquaculture industry, not just in finfish, has been thoroughly reviewed [126]. Below are some examples of research into bacteriophage applications in aquaculture, to demonstrate the breadth of the target species and pathogens.

Finfish

The bacteria *Flavobacterium psychrophilum* is the cause of both cold-water disease and rainbow trout fry syndrome in salmonids [127]. Castillo et al. isolated strains of *F. psychrophilum* and bacteriophages active against these bacteria from salmonid aquaculture sources. Six bacteriophages were characterized in detail and some resulted in a decrease in mortality from *F. psychrophilum* under laboratory conditions. The bacteriophages were administered at the same time as the fish were infected with the bacteria [128]. *F. psychrophilum* also causes cold-water disease in Japanese ayu (*Plecoglossus altivelis altivelis*) [129]. Five bacteriophages were tested for their ability to reduce the growth of strains of *P. psychrophilum* from ayu and other fish and were shown to be effective *in vitro* [130]. Stenholm et al. also isolated and characterized bacteriophages active against *F. psychrophilum* from rainbow trout farms. The bacteriophages had varying host ranges, with some able to infect up to 23 of 28 strains of *F. psychrophilum* tested [131].

Aeromonas salmonicida is another significant pathogen in finfish aquaculture. It causes furunculosis and bacterial septicaemia in a wide range of fish species [132]. Kim et al. isolated and characterized a lytic bacteriophage, from rainbow trout farm sediment samples, that had broad lytic activity against *A. salmonicida in vitro* [133]. The activity of a bacteriophage active against *A. salmonicida* has also been investigated in an aquarium situation for brook trout. Introduction of the bacteriophage into the aquariums resulted in a decrease in *A. salmonicida* [134]. Pereira et al. isolated separate bacteriophages with activity against *Vibrio parahaemolyticus* (*V. parahaemolyticus*) and *A. salmonicida* [135]. *V. parahaemolyticus* is one of a number of microorganisms in aquaculture species that pose a zoonotic risk to humans. It is a common cause of foodborne disease in humans acquired through ingestion of contaminated seafood [136, 137]. The bacteriophages isolated and characterized in this study were selected based on their lytic activity, survival time in marine water and specificity for the pathogenic bacteria species, to minimize the impact on the local bacterial community in the environment [135].

Another study, however, reported that furunculosis caused by *A. salmonicida* could not be adequately controlled via the use of bacteriophage therapy in Atlantic salmon. Mortality was delayed in the bacteriophage-treated fish, but the overall mortality was the same. Bacteriophage-resistant strains of *A. salmonicida* were also isolated [138]. *Edwardsiella ictaluri* (*Edw. ictaluri*) is another bacterial pathogen of finfish. It is a significant pathogen of farm-raised channel catfish, causing enteric septicaemia [139]. In one study, two bacteriophages were isolated with activity against *Edw. ictaluri*. These bacteriophages showed significant activity *in vitro* against *Edw. ictaluri* isolates [140].

Shellfish

Shellfish are another potential source of human foodborne infection. Human enteric viruses are of particular concern [141, 142]. In bivalve molluscan shellfish such as oysters, mussels and scallops, *E. coli* is commonly used as an indicator for the suitability of the shellfish for human consumption. However this is now considered a substandard method of assessment because human enteric viruses may still be present in shellfish that meet the standards for human consumption based on *E. coli* [143, 144]. Dore et al. evaluated an F-specific RNA (FRNA) bacteriophage as a potential indicator of human enteric viruses in bivalve molluscan shellfish [145].

FRNA bacteriophages are single-stranded viruses that can be present in shellfish as they are abundant in sewage, which is the main contaminant of filter-feeding bivalve molluscs [146, 147]. All of the shellfish tested in this study complied with standards for *E. coli* of less than 230 colony forming units (CFU) per 100 g of shellfish flesh. In comparison, FRNA bacteriophages were often detected in the same samples at levels exceeding 1,000 PFU per 100 g of flesh. There was a strong association demonstrated between high levels of FRNA bacteriophages, faecal pollution in the local harvest area and outbreaks of shell-fish associated disease [145]. This is one of numerous potential applications for bacteriophages outside of therapeutic use.

Crustaceans

Vibrio harveyi is one bacterial pathogen that can cause disease in cultured crustaceans [148]. A number of studies have reported the isolation of bacteriophages active against *V. harveyi*. One study isolated four bacteriophages, three from oyster tissue and one from shrimp hatchery water. The bacteriophages were capable of lysing between 55 and 70 percent of the *V. harveyi* isolates tested. Bacteriophage treatment was shown to decrease the mortality of shrimp in a hatchery environment due to *V. harveyi* [149]. Crothers-Stomps et al. also isolated and characterized lytic bacteriophages active against *V. harveyi* from discharge channels and grow-out ponds of a prawn farm. These bacteriophages had a broad host range including strains of *V. harveyi*, *V. campbellii*, *V. rotiferianus* and *V. parahaemolyticus* [150]. Another bacteriophage that can infect *V. harveyi* was isolated from shrimp farm water, which demonstrated a broad lytic ability for strains of *V. harveyi*. *Penaeus monodon* larvae infected with *V. harveyi* had an enhanced survival rate when in the presence of the bacteriophage compared to infected controls. In a field trial in a commercial hatchery the bacteriophage was successful in improving the survival of larvae in a naturally-occurring outbreak of *V. harveyi* infection. The bacteriophage also decreased the levels of *V. harveyi* in the tanks [151].

Food Products

The risk of transfer of zoonotic pathogens via the foodborne route is of paramount concern in products from animal sources [152, 153]. The first two bacteriophage products to be approved for use on live animals for the purposes of control of zoonotic bacteria were produced by OmniLytics™ and were targeted against *E. coli* O157:H7 and *Salmonella*, primarily as hide sprays or treatments for animal holding areas [43, 44]. The *Salmonella* bacteriophage products are approved for use in poultry, as a treatment to reduce the level of *Salmonella* on the feathers of the live birds prior to processing [154]. The time immediately prior to slaughter and processing is an important critical control point for intervention to mitigate the risk of significant levels of foodborne disease organisms ending up in food products.

Other bacteriophage products and or proposed control measures for bacterial contamination are designed to control bacteria directly in the food products themselves, with no involvement of treatment of live animals. One important pathogen that has been targeted by bacteriophage biocontrol strategies is *Listeria monocytogenes*. *L. monocytogenes* is a zoonotic pathogen which can be transmitted via the foodborne route [155]. It is a psychrotrophic organism, which means it is capable of survival and growth at low temperatures and therefore growth is not inhibited by refrigeration [156]. In 2006 the United States Food and Drug Administration approved the use of a bacteriophage product produced by Intralytix, Inc. (Baltimore, MD) designed to target *L. monocytogenes*. This was the first bacteriophage product approved in the United States for direct use in food. The product (ListShield™/LMP-102™) is designed for use on ready-to-eat meat and poultry products via a spray and is reported to be effective against at least 170 different strains of *L. monocytogenes* [15].

Another product approved for control of *L. monocytogenes* in foods in LISTEX™ P100, which is for use in the production of meat, cheese, fish, vegetables, fruit and other food products. It reportedly has a broad spectrum of activity against strains of *L. monocytogenes* [14]. The same company has more recently gained approval for a bacteriophage product designed for use against *Salmonella* (Salmonelex™) [157]. Numerous studies have reported the efficacy of these and other bacteriophage products against pathogens on various food products from animal sources [62, 158-161]. The use of bacteriophage-based biocontrol strategies to reduce foodborne pathogens in foods has been recently reviewed. This review discusses both pre- and post-harvest control of foodborne organisms [45]. An earlier review by Garcia et al. also discusses the application of bacteriophages in food safety [162].

Conclusion

Animal agriculture is a vital component of world food production. Adequate control of clinical and subclinical disease in farmed food animals is essential for efficient and sustainable production. All animals used in food production are also hosts of pathogens that can infect humans via foodborne transmission or other routes of infection. In considering ways to limit the risk of transmission of foodborne pathogens in products from animal sources there are numerous points in the chain of production where interventions are used. Bacteriophages have been tested at these different stages, including control in live animals on-farm, treatment of animals prior to slaughter and processing, and direct application of bacteriophages to food products.

The self-replicating and self-limiting nature of bacteriophages makes them an attractive potential alternative to other antimicrobials. They are naturally-occurring entities, meaning they are likely to have an advantage in markets where organic or chemical free products are desired. The fact that they have co-evolved with their bacterial hosts means that harnessing their inherent properties to completely eliminate their host bacteria is challenging. The emergence of products for use in foods, for direct application on animal hides or feathers, and treatment of animal holding environments, is an indicator of the commercial applications of bacteriophages but also the enormous challenges in using bacteriophages for direct treatment of animals. Administering bacteriophages to an animal on-farm for treatment or control of bacterial pathogens shows considerable promise, but the many studies that have been conducted also demonstrate the challenges that must be overcome in order to use bacteriophages for this purpose. This is not the only potential role for bacteriophages in the chain of food production, however. The products available for pre-processing usage are one alternative example. Bacteriophages have also been investigated as potential vaccine agents and for detection of foodborne pathogens, such as their use as an indicator of potential human enteric virus contamination in shellfish.

Antibiotic resistance of animal pathogens and zoonotic bacteria is of concern in the safe production of food. Therefore, suitable adjunctive methods are sought as alternatives for antibiotics in animal agriculture. Safe, natural, effective antimicrobials are highly desirable for use in many areas of food production, from direct animal treatment through to food applications. Research and the recent development of a number of bacteriophage products that have applications in production of food from animal sources suggests that bacteriophages are likely to play a significant future role in production of food from animal sources that is safe for human consumption.

References

[1] Economic Research Service - USDA. Animal Production and Marketing Issues. Economic Research Service - United States Department of Agriculture; 2012 [cited 2013 1st January]; Available from: http://www.ers.usda.gov/topics/animal-products/animal-production-marketing-issues.aspx.

[2] USDA. Cattle Inventory Report. United States: United States Department of Agriculture; 2012 [cited 2013 9th January]; Available from: http://usda01.library.cornell.edu/usda/nass/Catt//2010s/2012/Catt-01-27-2012.pdf.

[3] USDA. Sheep and goat review. United States Department of Agriculture - National Agricultural Statistics Service; 2012 [cited 2013 9th January]; Available from: http://www.nass.usda.gov/Statistics_by_State/Wisconsin/Publications/Livestock/sheep_goat_review.pdf.

[4] Australian Bureau of Statistics. Agricultural Commodities, Australia, 2010-11. Australian Bureau of Statistics; 2012 [cited 2013 9th January]; Available from: http://www.abs.gov.au/ausstats/abs@.nsf/Latestproducts/7121.0Main%20Features82010-11?opendocument&tabname=Summary&prodno=7121.0&issue=2010-11&num=&view=.

[5] United Nations. World population to reach 10 billion by 2100 if fertility in all countries converges to replacement level. United Nations; 2011 [cited 2013 9th January]; Available from: http://esa.un.org/wpp/Other-Information/Press_Release_WPP2010.pdf.

[6] Nelson JM, Chiller TM, Powers JH, Angulo FJ. Fluoroquinolone-Resistant Campylobacter Species and the Withdrawal of Fluoroquinolones from Use in Poultry: A Public Health Success Story. *Clinical Infectious Diseases*. 2007;44(7):977-80.

[7] Landers TF, Cohen B, Wittum TE, Larson EL. A review of antibiotic use in food animals: perspective, policy, and potential. *Public Health Reports*. 2012;127(1):4-22.

[8] Mirza F. Veterinary drug residues and MRLs. *Animal Pharm Reports*. Abingdon; UK: Informa Healthcare; 2006.

[9] OzFoodNet. Monitoring the incidence and causes of diseases potentially transmitted by food in Australia: Annual report of the OzFoodNet network, 2010. *Communicable Diseases Intelligence Quarterly Report*. 2012;36(3):E213.

[10] Gould LH, Nisler AL, Herman KM, Cole DJ, Williams IT, Mahon BE, et al. Surveillance for foodborne disease outbreaks – United States, 2008. *Morbidity and Mortality Weekly Report*. 2011;60(35):1197-202.

[11] Kiranmayi CB, Krishnaiah N, Mallika EN. Escherichia coli O157:H7 - An Emerging Pathogen in foods of Animal Origin. *Veterinary World*. 2010;3(8):382-9.

[12] Buchanan RL, Doyle MP. Foodborne disease significance of Escherichia coli O157:H7 and other enterohemorrhagic E. coli. *Food Technology*. 1997;51(10):69-76.

[13] Low JC, McKendrick IJ, McKechnie C, Fenlon D, Naylor SW, Currie C, et al. Rectal carriage of enterohemorrhagic Escherichia coli O157 in slaughtered cattle. *Applied and Environmental Microbiology*. 2005;71(1):93-7.

[14] Micreos Food Safety. Safety, security and peace of mind. 2012 [cited 2013 1st January]; Available from: http://www.micreosfoodsafety.com/en/profile-mission.aspx.

[15] Lang LH. FDA Approves Use of Bacteriophages to be Added to Meat and Poultry Products. *Gastroenterology*. 2006;131(5):1370.

[16] Lambertz ST, Nilsson C, Bradenmark A, Sylven S, Johansson A, Jansson LM, et al. Prevalence and level of Listeria monocytogenes in ready-to-eat foods in Sweden 2010. *International Journal of Food Microbiology*. 2012;160(1):24-31.

[17] Kramarenko T, Roasto M, Meremae K, Kuningas M, Poltsama P, Elias T. Listeria monocytogenes prevalence and serotype diversity in various foods. *Food Control*. 2013;30(1):24-9.

[18] Hanlon GW. Bacteriophages: an appraisal of their role in the treatment of bacterial infections. *International Journal of Antimicrobial Agents*. 2007;30(2):118-28.

[19] Thompson GD. Consumer Demand for Organic Foods: What We Know and What We Need to Know. *American Journal of Agricultural Economics*. 1998(5):1113-1118.

[20] Callaway TR, Edrington TS, Brabban AD, Keen JE, Anderson RC, Rossman ML, et al. Fecal prevalence of Escherichia coli O157, Salmonella, Listeria, and bacteriophage infecting E. coli O157:H7 in feedlot cattle in the Southern Plains Region of the United States. *Foodborne Pathogens and Disease*. 2006;3(3):234-44.

[21] Berg Miller ME, Yeoman CJ, Chia N, Tringe SG, Angly FE, Edwards RA, et al. Phage-bacteria relationships and CRISPR elements revealed by a metagenomic survey of the rumen microbiome. *Environmental Microbiology*. 2012;14(1):207-27.

[22] Alderman DJ, Hastings TS. Antibiotic use in aquaculture: development of antibiotic resistance – potential for consumer health risks. *International Journal of Food Science and Technology*. 1998;33(2):139-55.

[23] Brussow H, Canchaya C, Hardt WD. Phages and the evolution of bacterial pathogens: From genomic rearrangements to lysogenic conversion. *Microbiology and Molecular Biology Reviews*. 2004;68(3):560-602.

[24] Smith DL, Rooks DJ, Fogg PCM, Darby AC, Thomson NR, McCarthy AJ, et al. Comparative genomics of Shiga toxin encoding bacteriophages. *BMC Genomics*. 2012;13(1):311-20.

[25] Waldor MK, Mekalanos JJ. Lysogenic Conversion by a Filamentous Phage Encoding Cholera Toxin. *Science*. 1996(5270):1910-1914.

[26] Ahn SI, Azzouny RA, Huyen TTT, Kwak HS. The characteristics, detection and control of bacteriophage in fermented dairy products. *Korean Journal for Food Science of Animal Resources*. 2009;29(1):1-14.

[27] Bouwman AF, Van der Hoek KW, Eickhout B, Soenario I. Exploring changes in world ruminant production systems. *Agricultural Systems*. 2005;84(2):121-53.

[28] Meat and Livestock Australia (MLA). Beef. Meat and Livestock Australia; 2013 [cited 2013 9th January]; Available from: http://www.mla.com.au/Prices-and-markets/Overseas-markets/North-America/Beef.

[29] Dairy Australia. Dairy production and trade. Dairy Australia; 2013 [cited 2013 9th January]; Available from: http://www.dairyaustralia.com.au/Statistics-and-markets/Exports-and-trade/Dairy-Trade-Environment/Dairy-Production-and-Trade.aspx.

[30] Callaway TR, Carr MA, Edrington TS, Anderson RC, Nisbet DJ. Diet, Escherichia coli O157:H7, and cattle: a review after 10 years. *Current Issues In Molecular Biology*. 2009;11(2):67-79.

[31] O'Brien SB, Duffy G, Carney E, Sheridan JJ, McDowell DA, Blair IS. Prevalence and numbers of Escherichia coli O157 on bovine hides at a beef slaughter plant. *Journal of Food Protection*. 2005;68(4):660-5.

[32] Teunis P, Takumi K, Shinagawa K. Dose response for infection by Escherichia coli O157:H7 from outbreak data. *Risk Analysis*. 2004;24(2):401-7.

[33] Lanjewar M, De A, Mathur M. Diarrheagenic E. coli in hospitalized patients: Special reference to Shiga-like toxin producing Escherichia coli. *Indian Journal of Pathology and Microbiology*. 2010(1):75.

[34] Niu YD, McAllister TA, Xu Y, Johnson RP, Stephens TP, Stanford K. Prevalence and Impact of Bacteriophages on the Presence of Escherichia coli O157: H7 in Feedlot Cattle and Their Environment. *Applied and Environmental Microbiology*. 2009;75(5):1271-8.

[35] Niu YD, Johnson RP, Xu Y, McAllister TA, Sharma R, Louie M, et al. Host range and lytic capability of four bacteriophages against bovine and clinical human isolates of Shiga toxin-producing Escherichia coli O157:H7. *Journal of Applied Microbiology*. 2009;107(2):646-56.

[36] Rozema EA, Stephens TP, Bach SJ, Okine EK, Johnson RP, Stanford K, et al. Oral and rectal administration of bacteriophages for control of Escherichia coli O157:H7 in feedlot cattle. *Journal of Food Protection*. 2009;72(2):241-50.

[37] Stanford K, McAllister TA, Niu YD, Stephens TP, Mazzocco A, Waddell TE, et al. Oral delivery systems for encapsulated bacteriophages targeted at Escherichia coli O157:H7 in feedlot cattle. *Journal of Food Protection*. 2010;73(7):1304-12.

[38] Dini C, de Urraza PJ. Isolation and selection of coliphages as potential biocontrol agents of enterohemorrhagic and Shiga toxin-producing E. coli(EHEC and STEC) in cattle. *Journal of Applied Microbiology*. 2010;109(3):873-87.
[39] Oppenheim AB, Kobiler O, Stavans J, Court DL, Adhya S. Switches in bacteriophage lambda development. *Annual Review Of Genetics*. 2005;39:409-29.
[40] Coffey B, Rivas L, Duffy G, Coffey A, Ross RP, McAuliffe O. Assessment of Escherichia coli O157:H7-specific bacteriophages e11/2 and e4/1c in model broth and hide environments. *International Journal of Food Microbiology*. 2011;147(3):188-94.
[41] Rivas L, Coffey B, McAuliffe O, McDonnell MJ, Burgess CM, Coffey A, et al. In Vivo and Ex Vivo Evaluations of Bacteriophages e11/2 and e4/1c for Use in the Control of Escherichia coli O157:H7. *Applied and Environmental Microbiology*. 2010;76(21):7210-6.
[42] O'Flynn G, Ross RP, Fitzgerald GF, Coffey A. Evaluation of a cocktail of three bacteriophages for biocontrol of Escherichia coli O157:H7. *Applied and Environmental Microbiology*. 2004;70(6):3417-3424.
[43] OmniLytics. OmniLytics Announces USDA/FSIS Approval for Bacteriophage Treatment of E. coli O157:H7 on Livestock. 2007 [cited 2013 1st January]; Available from: http://www.businesswire.com/news/home/20070102005459/en/OmniLytics-Announces-USDAFSIS-Approval-Bacteriophage-Treatment-E.
[44] OmniLytics. OmniLytics Announces USDA/FSIS Allowance of Bacteriophage Treatment of Salmonella on Livestock. 2007 [cited 2013 1st January]; Available from: http://www.businesswire.com/news/home/20070329005872/en/OmniLytics-Announces-USDAFSIS-Allowance-Bacteriophage-Treatment-Salmonella.
[45] Goodridge LD, Bisha B. Phage-based biocontrol strategies to reduce foodborne pathogens in foods. *Bacteriophage*. 2011;1(3):130-7.
[46] Constable PD. Antimicrobial use in the treatment of calf diarrhea. *Journal of Veterinary Internal Medicine*. 2004;18(1):8-17.
[47] Bicalho MLS, Machado VS, Nydam DV, Santos TMA, Bicalho RC. Evaluation of oral administration of bacteriophages to neonatal calves: Phage survival and impact on fecal Escherichia coli. *Livestock Science*. 2012;144(3):294-9.
[48] Bradley AJ. Bovine Mastitis: An Evolving Disease. *The Veterinary Journal*. 2002;164(2):116-28.
[49] Anderson KL, Lyman R, Moury K, Ray D, Watson DW, Correa MT. Molecular epidemiology of Staphylococcus aureus mastitis in dairy heifers. *Journal Of Dairy Science*. 2012;95(9):4921-30.
[50] Costa GM, Paiva LV, Figueiredo HCP, Figueira AR, Pereira UP, Silva N. Population diversity of Staphylococcus aureus isolated from bovine mastitis in Brazilian dairy herds. *Research In Veterinary Science*. 2012;93(2):733-5.
[51] O'Flaherty S, Ross RP, Flynn J, Meaney WJ, Fitzgerald GF, Coffey A. Isolation and characterization of two anti-staphylococcal bacteriophages specific for pathogenic Staphylococcus aureus associated with bovine infections. *Letters in Applied Microbiology*. 2005;41(6):482-6.
[52] Kwiatek M, Parasion S, Mizak L, Gryko R, Bartoszcze M, Kocik J. Characterization of a bacteriophage, isolated from a cow with mastitis, that is lytic against Staphylococcus aureus strains. *Archives of Virology*. 2012;157(2):225-34.

[53] Donovan DM, Lardeo M, Foster-Frey J. Lysis of staphylococcal mastitis pathogens by bacteriophage phi11 endolysin. *FEMS Microbiology Letters.* 2006;265(1):133-9.

[54] Gill JJ, Pacan JC, Carson ME, Leslie KE, Griffiths MW, Sabour PM. Efficacy and pharmacokinetics of bacteriophage therapy in treatment of subclinical Staphylococcus aureus mastitis in lactating dairy cattle. *Antimicrobial Agents and Chemotherapy.* 2006;50(9):2912-8.

[55] Sheldon I. The postpartum uterus. Veterinary Clinics of North America: *Food Animal Practice.* 2004;20:569-91.

[56] Bicalho RC, Santos TMA, Gilbert RO, Caixeta LS, Teixeira LM, Bicalho MLS, et al. Susceptibility of Escherichia coli isolated from uteri of postpartum dairy cows to antibiotic and environmental bacteriophages. Part I: Isolation and lytic activity estimation of bacteriophages. *Journal Of Dairy Science.* 2010;93(1):93-104.

[57] Santos TM, Gilbert RO, Caixeta LS, Machado VS, Teixeira LM, Bicalho RC. Susceptibility of Escherichia coli isolated from uteri of postpartum dairy cows to antibiotic and environmental bacteriophages. Part II: In vitro antimicrobial activity evaluation of a bacteriophage cocktail and several antibiotics. *Journal Of Dairy Science.* 2010;93(1):105-14.

[58] Machado VS, Bicalho MLS, Pereira RV, Caixeta LS, Bittar JHJ, Oikonomou G, et al. The effect of intrauterine administration of mannose or bacteriophage on uterine health and fertility of dairy cows with special focus on Escherichia coli and Arcanobacterium pyogenes. *Journal Of Dairy Science.* 2012;95:3100-9.

[59] Santorum P, Garcia R, Lopez V, Martinez-Suarez JV. Dairy farm management and production practices associated with the presence of Listeria monocytogenes in raw milk and beef. *Spanish Journal of Agricultural Research.* 2012;10(2):360-71.

[60] Vongkamjan K, Switt AM, Bakker HCd, Fortes ED, Wiedmann M. Silage collected from dairy farms harbors an abundance of listeriaphages with considerable host range and genome size diversity. *Applied and Environmental Microbiology.* 2012;78(24):8666-75.

[61] Modi R, Hirvi Y, Hill A, Griffiths MW. Effect of phage on survival of Salmonella Enteritidis during manufacture and storage of Cheddar cheese made from raw and pasteurized milk. *Journal of Food Protection.* 2001;64(7):927-33.

[62] Carlton RM, Noordman WH, Biswas B, de Meester ED, Loessner MJ. Bacteriophage P100 for control of Listeria monocytogenes in foods: Genome sequence, bioinformatic analyses, oral toxicity study, and application. *Regulatory Toxicology and Pharmacology.* 2005;43(3):301-12.

[63] García P, Madera C, Martínez B, Rodríguez A. Biocontrol of Staphylococcus aureus in curd manufacturing processes using bacteriophages. *International Dairy Journal.* 2007;17(10):1232-9.

[64] García P, Madera C, martínez B, Rodríguez A, Suárez JE. Prevalence of bacteriophages infecting Staphylococcus aureus in dairy samples and their potential as biocontrol agents. *Journal Of Dairy Science.* 2009;92(7):3019-26.

[65] Mahony J, Ainsworth S, Stockdale S, van Sinderen D. Phages of lactic acid bacteria: The role of genetics in understanding phage-host interactions and their co-evolutionary processes. *Virology.* 2012;434(2):143-50.

[66] Carminati D, Giraffa G, Quiberoni A, Binetti A, Suárez V, Reinheimer J. Advances and trends in starter cultures for dairy fermentations. In: Mozzi F, Raya R, Vignolo G,

editors. Biotechnology of lactic acid bacteria: Novel applications. Iowa, USA: Wiley-Blackwell; 2010. 177-92.

[67] Brüssow H. Phages of Dairy Bacteria. *Annual Review of Microbiology*. 2001;55(1):283-303.

[68] Marcó MB, Moineau S, Quiberoni A. Bacteriophages and dairy fermentations. *Bacteriophage*. 2012;2(3):1-10.

[69] Iverson WG, Millis NF. Succession of Streptococcus bovis strains with differing bacteriophage sensitivities in the rumens of two fistulated sheep. *Applied and Environmental Microbiology*. 1977;33(4):810-3.

[70] Swain RA, Nolan JV, Klieve AV. Natural variability and diurnal fluctuations within the bacteriophage population of the rumen. *Applied and Environmental Microbiology*. *1*996;62(3):994-7.

[71] Cheong JPE, Brooker JD. Isolation of a virulent bacteriophage from a Propionibacterium species in the sheep rumen. *Australian Journal of Agricultural Research*. 2000;51(1):119-23.

[72] Klieve AV, Swain RA. Estimation of ruminal bacteriophage numbers by pulsed-field gel-electrophoresis and laser densitometry. *Applied and Environmental Microbiology*. 1993;59(7):2299-303.

[73] Koch C, Hertwig S, Lurz R, Appel B, Beutin L. Isolation of a lysogenic bacteriophage carrying the stx1OX3 gene, which is closely associated with Shiga toxin-producing Escherichia coli strains from sheep and humans. *Journal of Clinical Microbiology*. 2001;39(11):3992-8.

[74] Döpfer D, Sekse C, Beutin L, Solheim H, van der Wal FJ, de Boer A, et al. Pathogenic potential and horizontal gene transfer in ovine gastrointestinal Escherichia coli. *Journal of Applied Microbiology*. 2010;108(5):1552-62.

[75] Brandal LT, Sekse C, Lindstedt B-A, Sunde M, Løbersli I, Urdahl AM, et al. Norwegian Sheep Are an Important Reservoir for Human-Pathogenic Escherichia coli O26:H11. *Applied and Environmental Microbiology*. 2012;78(12):4083-91.

[76] Cornick NA, Helgerson AF, Sharma V. Shiga Toxin and Shiga Toxin-Encoding Phage Do Not Facilitate Escherichia coli O157:H7 Colonization in Sheep. *Applied and Environmental Microbiology*. 2007;73(1):344-6.

[77] Bach SJ, McAllister TA, Veira DM, Gannon VPJ, Holley RA. Effect of bacteriophage DC22 on Escherichia coli O157 : H7 in an artificial rumen system (Rusitec) and inoculated sheep. *Animal Research*. 2003;52(2):89-101.

[78] Raya RR, Varey P, Oot RA, Dyen MR, Callaway TR, Edrington TS, et al. Isolation and Characterization of a New T-Even Bacteriophage, CEV1, and Determination of Its Potential To Reduce Escherichia coli O157:H7 Levels in Sheep. *Applied and Environmental Microbiology*. 2006;72(9):6405-10.

[79] Wang S, Zhao W, Raza A, Friendship R, Johnson R, Kostrzynska M, et al. Prevalence of Salmonella infecting bacteriophages associated with Ontario pig farms and the holding area of a high capacity pork processing facility. *Journal of the Science of Food and Agriculture*. 2010;90(13):2318-25.

[80] Larsen ST, Hurd HS, McKean JD, Griffith RW, Wesley IV. Effect of short-term lairage on the prevalence of Salmonella enterica in cull sows. *Journal of Food Protection*. 2004;67(7):1489-93.

[81] Hurd HS, McKean JD, Griffith RW, Wesley IV, Rostagno MH. Salmonella enterica infections in market swine with and without transport and holding. *Applied and Environmental Microbiology*. 2002;68(5):2376-81.

[82] Wall SK, Jiayi Z, Rostagno MH, Ebner PD. Phage Therapy To Reduce Preprocessing Salmonella Infections in Market-Weight Swine. *Applied and Environmental Microbiology*. 2010;76(1):48-53.

[83] Saez AC, Zhang JY, Rostagno MH, Ebner PD. Direct Feeding of Microencapsulated Bacteriophages to Reduce Salmonella Colonization in Pigs. *Foodborne Pathog Dis*. 2011;8(12):1269-74.

[84] Hooton SPT, Atterbury RJ, Connerton IF. Application of a bacteriophage cocktail to reduce Salmonella Typhimurium U288 contamination on pig skin. *International Journal of Food Microbiology*. 2011;151(2):157-63.

[85] Jamalludeen N, Johnson RP, Shewen PE, Gyles CL. Evaluation of bacteriophages for prevention and treatment of diarrhea due to experimental enterotoxigenic Escherichia coli O149 infection of pigs. *Veterinary Microbiology*. 2009;136(1–2):135-41.

[86] Cox E, Schrauwen E, Cools V, Houvenaghel A. Experimental Induction of Diarrhea in Newly-Weaned Piglets. *Journal of Veterinary Medicine Series A*. 1991;38(6):418-26.

[87] Cha SB, Yoo AN, Lee WJ, Shin MK, Jung MH, Shin SW, et al. Effect of Bacteriophage in Enterotoxigenic Escherichia coli (ETEC) Infected Pigs. *Journal of Veterinary Medical Science*. 2012;74(8):1037-9.

[88] Manoutcharian K, Díaz-Orea A, Gevorkian G, Fragoso G, Acero G, González E, et al. Recombinant bacteriophage-based multiepitope vaccine against Taenia solium pig cysticercosis. *Veterinary Immunology and Immunopathology*. 2004;99(1–2):11-24.

[89] Ren ZJ, Tian CJ, Zhu QS, Zhao MY, Xin AG, Nie WX, et al. Orally delivered foot-and-mouth disease virus capsid protomer vaccine displayed on T4 bacteriophage surface: 100% protection from potency challenge in mice. *Vaccine*. 2008;26(11):1471-81.

[90] Shi YB, Li N, Yan YX, Wang HG, Li Y, Lu CP, et al. Combined Antibacterial Activity of Phage Lytic Proteins Holin and Lysin from Streptococcus suis Bacteriophage SMP. *Current Microbiology*. 2012;65(1):28-34.

[91] Zehr ES, Tabatabai LB, Bayles DO. Genomic and proteomic characterization of SuMu, a Mu-like bacteriophage infecting Haemophilus parasuis. *BMC Genomics*. 2012;13(1):331-43.

[92] Yan L, Hong SM, Kim IH. Effect of Bacteriophage Supplementation on the Growth Performance, Nutrient Digestibility, Blood Characteristics, and Fecal Microbial Shedding in Growing Pigs. *Asian-Australasian Journal of Animal Sciences*. 2012;25(10):1451-6.

[93] Hermans D, Pasmans F, Messens W, Martel A, Immerseel Fv, Rasschaert G, et al. Poultry as a host for the zoonotic pathogen Campylobacter jejuni. *Vector Borne and Zoonotic Diseases*. 2012;12(2):89-98.

[94] Atterbury RJ, Connerton PL, Dodd CER, Rees CED, Connerton IF. Isolation and Characterization of Campylobacter Bacteriophages from Retail Poultry. *Applied and Environmental Microbiology*. 2003;69(8):4511-8.

[95] Carvalho CM, Gannon BW, Halfhide DE, Santos SB, Hayes CM, Roe JM, et al. The in vivo efficacy of two administration routes of a phage cocktail to reduce numbers of Campylobacter coli and Campylobacter jejuni inchickens. *BMC Microbiology*. 2010;10:232-42.

[96] El-Shibiny A, Scott A, Timms A, Metawea Y, Connerton P, Connerton I. Application of a group II Campylobacter bacteriophage to reduce strains of Campylobacter jejuni and Campylobacter coli colonizing broiler chickens. *Journal of Food Protection*. 2009;72(4):733-40.

[97] Scott AE, Timms AR, Connerton PL, Loc Carrillo C, Adzfa Radzum K, Connerton IF. Genome dynamics of Campylobacter jejuni in response to bacteriophage predation. *Plos Pathogens*. 2007;3(8):e119-e.

[98] Owens J, Barton MD, Heuzenroeder MW. The isolation and characterization of Campylobacter jejuni bacteriophages from free range and indoor poultry. *Veterinary Microbiology*. 2013;162: 144-150.

[99] Connerton PL, Timms AR, Connerton IF. Campylobacter bacteriophages and bacteriophage therapy. *Journal of Applied Microbiology*. 2011;111(2):255-65.

[100] Zhao S, Maurer JJ, Hubert S, De Villena JF, McDermott PF, Meng J, et al. Antimicrobial susceptibility and molecular characterization of avian pathogenic Escherichia coli isolates. *Veterinary Microbiology*. 2005;107(3–4):215-24.

[101] Barnes HJ, Gross WB. Collibacillosis. In: Calnek BW, Barnes HJ, Beard CW, McDougald LR, Saif YM, editors. Diseases of Poultry. Ames, IA: Iowa State University Press; 1997. 131-41.

[102] Oliveira A, Sillankorva S, Quinta R, Henriques A, Sereno R, Azeredo J. Isolation and characterization of bacteriophages for avian pathogenic E. coli strains. *Journal of Applied Microbiology*. 2009;106(6):1919-27.

[103] Oliveira A, Sereno R, Azeredo J. In vivo efficiency evaluation of a phage cocktail in controlling severe colibacillosis in confined conditions and experimental poultry houses. *Veterinary Microbiology*. 2010;146 (3–4):303-8.

[104] Jamalludeen N, She YM, Lingohr EJ, Griffiths M. Isolation and characterization of virulent bacteriophages against Escherichia coli serogroups O1, O2, and O78. *Poultry Science*. 2009;88(8):1694-702.

[105] Barrow P, Lovell M, Berchieri A, Jr. Use of lytic bacteriophage for control of experimental Escherichia coli septicemia and meningitis in chickens and calves. *Clinical And Diagnostic Laboratory Immunology*. 1998;5(3):294-8.

[106] Lau GL, Sieo CC, Tan WS, Hair-Bejo M, Jalila A, Ho YW. Efficacy of a bacteriophage isolated from chickens as a therapeutic agent for colibacillosis in broiler chickens. *Poultry Science*. 2010;89(12):2589-96.

[107] Huff WE, Huff GR, Rath NC, Balog JM, Donoghue AM. Therapeutic efficacy of bacteriophage and Baytril (enrofloxacin) individually and in combination to treat colibacillosis in broilers. *Poultry Science*. 2004;83(12):1944-7.

[108] Huff WE, Huff GR, Rath NC, Donoghue AM. Immune interference of bacteriophage efficacy when treating colibacillosis in poultry. *Poultry Science*. 2010;89(5):895-900.

[109] Duffy LL, Dykes GA, Fegan N. A review of the ecology, colonization and genetic characterization of Salmonella enterica serovar Sofia, a prolific but avirulent poultry serovar in Australia. *Food Research International*. 2012;45(2):770-9.

[110] Fiorentin L, Vieira N, Barioni Jr W. Oral treatment with bacteriophages reduces the concentration of Salmonella Enteritidis PT4 in caecal contents of broilers. *Avian Pathology*. 2005;34(3):258-63.

[111] Fiorentin LF, Vieira ND, Barioni Junior W, Barros S. 'In vivo' characterization and 'in vivo' properties of Salmonellae lytic bacteriophages isolated from free-range layers. *Brazilian Journal of Poultry Science.* 2004;6(2):121-128.
[112] Chung YH, Kim SY, Chang YH. Prevalence and antibiotic susceptibility of Salmonella isolated from foods in Korea from 1993 to 2001. *Journal of Food Protection.* 2003;66(7):1154-7.
[113] Atterbury RJ, Van Bergen MAP, Ortiz F, Lovell MA, Harris JA, De Boer A, et al. Bacteriophage therapy to reduce Salmonella colonization of broiler chickens. Applied and Environmental Microbiology. 2007;73(14):4543-9.
[114] Hurley A, Maurer JJ, Lee MD. Using bacteriophages to modulate Salmonella colonization of the chicken's gastrointestinal tract: lessons learned from in silico and in vivo modeling. *Avian Diseases*. 2008;52(4):599-607.
[115] Robeson J, Valencia M, Retamales J, Borie C. Stability inside hen eggs of a Salmonella enterica serovar Enteritidis bacteriophage. *Electronic Journal of Biotechnology.* 2011;14(4):1-4.
[116] Prakash B, Krishnappa G, Muniyappa L, Kumar BS. Epidemiological Characterization of Avian Salmonella enterica Serovar Infections in India. *International Journal of Poultry Science.* 2005(6):388.
[117] Orji MU, Onuigbo HC, Mbata TI. Isolation of Salmonella from poultry droppings and other environmental sources in Awka, Nigeria. *International Journal of Infectious Diseases.* 2005;9(2):86-9.
[118] Bao H, Zhang H, Wang R. Isolation and characterization of bacteriophages of Salmonella enterica serovar Pullorum. *Poultry Science*. 2011;90(10):2370-7.
[119] Kwon YK, Kim A, Kang MS, Her M, Jung BY, Lee KM, et al. Prevalence and characterization of Salmonella Gallinarum in the chicken in Korea during 2000 to 2008. *Poultry Science.* 2010;89(2):236-42.
[120] Lee Y, Kim K, Kwon Y, Kang M, Mo I, Kim J, et al. Prevalent characteristics of fowl typhoid in Korea. *Journal of Veterinary Clinics*. 2003;20(2):155-8.
[121] Lim T-H, Lee D-H, Lee Y-N, Park J-K, Youn H-N, Kim M-S, et al. Efficacy of bacteriophage therapy on horizontal transmission of Salmonella gallinarum on commercial layer chickens. *Avian Diseases*. 2011;55(3):435-8.
[122] Miller RW, Skinner EJ, Sulakvelidze A, Mathis GF, Hofacre CL. Bacteriophage therapy for control of necrotic enteritis of broiler chickens experimentally infected with Clostridium perfringens. *Avian Diseases.* 2010;54(1):33-40.
[123] Helmboldt CF, Bryant ES. The Pathology of Necrotic Enteritis in Domestic Fowl. *Avian Diseases*. 1971(4):775.
[124] Cabello FC. Heavy use of prophylactic antibiotics in aquaculture: a growing problem for human and animal health and for the environment. *Environmental Microbiology.* 2006;8(7):1137-44.
[125] Nakai T, Park SC. Bacteriophage therapy of infectious diseases in aquaculture. *Research in Microbiology.* 2002;153(1):13-8.
[126] Oliveira J, Castilho F, Cunha A, Pereira MJ. Bacteriophage therapy as a bacterial control strategy in aquaculture. *Aquaculture International*. 2012;20(5):879-910.
[127] Nematollahi A, Decostere A, Pasmans F, Haesebrouck F. Review Flavobacterium psychrophilum infections in salmonid fish. *Journal of Fish Diseases.* 2003;26(10):563.

[128] Castillo D, Higuera G, Villa M, Middelboe M, Dalsgaard I, Madsen L, et al. Diversity of Flavobacterium psychrophilum and the potential use of its phages for protection against bacterial cold water disease in salmonids. *Journal of Fish Diseases*. 2012;35(3):193-201.

[129] Horiuchi M, Bunya T, Hoshiai G. Outbreaks of cold-water disease in coho salmon in Japan. *Fish Pathology*. 1991;26(4):211-2.

[130] Kim JH, Gomez DK, Nakai T, Park SC. Isolation and identification of bacteriophages infecting ayu Plecoglossus altivelis altivelis specific Flavobacterium psychrophilum. *Veterinary Microbiology*. 2010;140(1–2):109-15.

[131] Stenholm AR, Dalsgaard I, Middelboe M. Isolation and characterization of bacteriophages infecting the fish pathogen Flavobacterium psychrophilum. *Applied and Environmental Microbiology*. 2008;74(13):4070-8.

[132] Wiklund T, Dalsgaard I. Occurrence and significance of atypical Aeromonas salmonicida in non-salmonid and salmonid fish species: a review. *Diseases of Aquatic Organisms*. 1998;32(1):49-69.

[133] Kim J, Son J, Choi Y, Choresca C, Shin S, Han J, et al. Isolation and Characterization of a Lytic Myoviridae Bacteriophage PAS-1 with Broad Infectivity in Aeromonas salmonicida. *Current Microbiology*. 2012;64(5):418-26.

[134] Imbeault S, Parent S, Lagace M, Uhland CF, Blais JF. Using Bacteriophages to prevent furunculosis caused by Aeromonas salmonicida in farmed brook trout. *Journal of Aquatic Animal Health*. 2006;18(3):203-14.

[135] Pereira C, Silva YJ, Santos AL, Cunha Â, Gomes NCM, Almeida A. Bacteriophages with Potential for Inactivation of Fish Pathogenic Bacteria: Survival, Host Specificity and Effect on Bacterial Community Structure. *Marine Drugs*. 2011;9(11):2236-55.

[136] DePaola A, Ulaszek J, Kaysner CA, Tenge BJ, Nordstrom JL, Wells J, et al. Molecular, serological, and virulence characteristics of Vibrio parahaemolyticus isolated from environmental, food, and clinical sources in North America and Asia. *Applied and Environmental Microbiology*. 2003;69(7):3999-4005.

[137] Wong HC, Liu SH, Ku LW, Lee IY, Wang TK, Lee YS, et al. Characterization of Vibrio parahaemolyticus isolates obtained from foodborne illness outbreaks during 1992 through 1995 in Taiwan. *Journal of Food Protection*. 2000;63(7):900-6.

[138] Verner–Jeffreys DW, Algoet M, Pond MJ, Virdee HK, Bagwell NJ, Roberts EG. Furunculosis in Atlantic salmon (Salmo salar L.) is not readily controllable by bacteriophage therapy. *Aquaculture*. 2007;270(1–4):475-84.

[139] Hawke JP, McWhorter AC, Steigerwalt AG, Brenner DJ. Edwardsiella ictaluri sp. nov., the causative agent of enteric septicemia of catfish. *International Journal of Systematic Bacteriology*. 1981;31(4):396-400.

[140] Walakira JK, Carrias AA, Hossain MJ, Jones E, Terhune JS, Liles MR. Identification and characterization of bacteriophages specific to the catfish pathogen, Edwardsiella ictaluri. *Journal of Applied Microbiology*. 2008;105(6):2133-42.

[141] Guyader Fl, Haugarreau L, Miossec L, Dubois E, Pommepuy M. Three-year study to assess human enteric viruses in shellfish. *Applied and Environmental Microbiology*. 2000;66(8):3241-8.

[142] Boxman ILA. Human enteric viruses occurrence in shellfish from European markets. *Food and Environmental Virology*. 2010;2(3):156-66.

[143] West PA, Wood PC, Jacob M. Control of food poisoning risks associated with shellfish. *Journal of the Royal Society of Health*. 1985;105(1):15-21.

[144] Chalmers JWT, McMillan JH. An outbreak of viral gastroenteritis associated with adequately prepared oysters. *Epidemiology and Infection*. 1995;115(1):163-7.

[145] Dore WJ, Henshilwood K, Lees DN. Evaluation of F-specific RNA bacteriophage as a candidate human enteric virus indicator for bivalve molluscan shellfish. *Applied and Environmental Microbiology*. 2000;66(4):1280-5.

[146] Furuse K. Distribution of coliphages in the environment: general considerations. In: Goyal SM, Gerba CP, Bitton G, editor. Phage Ecology. New York: Wiley-Interscience; 1987:87-123.

[147] Havelaar AH, Olphen Mv, Drost YC. F-specific RNA bacteriophages are adequate model organisms for enteric viruses in fresh water. *Applied and Environmental Microbiology*. 1993;59(9):2956-62.

[148] Canak Ö, Yardmc RE. A review on the pathogen of cultured marine species: Vibrio harveyi. *Journal of FisheriesSciencescom*. 2011;5(2):131-40.

[149] Karunasagar I, Shivu MM, Girisha SK, Krohne G, Karunasagar I. Biocontrol of pathogens in shrimp hatcheries using bacteriophages. *Aquaculture*. 2007;268(1–4): 288-92.

[150] Crothers-Stomps C, Høj L, Bourne DG, Hall MR, Owens L. Isolation of lytic bacteriophage against Vibrio harveyi. *Journal of Applied Microbiology*. 2010;108(5):1744-50.

[151] Vinod MG, Shivu MM, Umesha KR, Rajeeva BC, Krohne G, Karunasagar I, et al. Isolation of Vibrio harveyi bacteriophage with a potential for biocontrol of luminous vibriosis in hatchery environments. *Aquaculture*. 2006;255(1–4):117-24.

[152] Köfer J, Schobesberger H. Animal hygiene and sustainable livestock production. Proceedings of the XVth International Congress of the International Society for Animal Hygiene, Vienna, Austria, 3-7 July 2011, Volume 2. Brno; Czech Republic: Tribun EU; 2011:599-1017.

[153] European Food Safety Authority. The European Union Summary Report on Trends and Sources of Zoonoses, Zoonotic Agents and Food-borne Outbreaks in 2010. *EFSA Journal*. 2012;10(3):2597-.

[154] OmniLytics. OmniLytics Announces USDA/FSIS Allowance of Bacteriophage Treatment of Salmonella on Poultry. OmniLytics; 2008 [cited 2013 1st January]; Available from: http://www.omnilytics.com/news/news021.html.

[155] Kramarenko T, Roasto M, Meremäe K, Kuningas M, Põltsama P, Elias T. Listeria monocytogenes prevalence and serotype diversity in various foods. *Food Control*. 2013;30(1):24-9.

[156] Walker SJ, Archer P, Banks JG. Growth of Listeria monocytogenes at refrigeration temperatures. *Journal of Applied Microbiology*. 1990;68(2):157-62.

[157] Micreos Food Safety. SALMONELEX^{TM} approved. Micreos; 2011 [cited 2013 1st January]; Available from: http://www.micreos.com/news/salmonelex-approved.aspx.

[158] Bigot B, Lee WJ, McIntyre L, Wilson T, Hudson JA, Billington C, et al. Control of Listeria monocytogenes growth in a ready-to-eat poultry product using a bacteriophage. *Food Microbiology*. 2011;28(8):1448-52.

[159] Soni KA, Nannapaneni R. Bacteriophage significantly reduces Listeria monocytogenes on raw salmon fillet tissue. *Journal of Food Protection*. 2010;73(1):32-8.

[160] Soni KA, Ramakrishna N, Hagens S. Reduction of Listeria monocytogenes on the surface of fresh channel catfish fillets by bacteriophage Listex p100. *Foodborne Pathogens and Disease*. 2010;7(4):427-34.

[161] Schellekens MM, Wouters J, Hagens S, Hugenholtz J. Bacteriophage P100 application to control Listeria monocytogenes on smeared cheese. *Milchwissenschaft – Milk Science International*. 2007;62(3):284-7.

[162] García P, Martínez B, Obeso JM, Rodríguez A. Bacteriophages and their application in food safety. *Letters in Applied Microbiology*. 2008;47(6):479-85.

In: Bacteriophages
Editors: Clark Denton and Richard J. Crosby
ISBN: 978-1-62618-513-5

Chapter 2

The Use of Phage: Therapy, Biocontrol and Commercial Microbiology

Roberto S. Dias, Monique R. Eller, Rafael L. Salgado, Cynthia C. da Silva and Sérgio O. De Paula[*]
Laboratory of Molecular Immunovirology, Department of General Biology, Federal University of Viçosa, Viçosa, Minas Gerais, Brazil

Abstract

The first studies involving phage dating from the early nineteenth century. In the last three decades these studies were intensified, mainly because of the emergence of antibiotic-resistant bacterial strains. Since then, the wide and diversity of phage species became evident. It is estimated that there are 10^{31} particles and 10^{8} different species of phages dispersed in the biosphere, which highlights their importance in ecosystems on the planet. These data generates the need for an efficient phage classification system. Classically, the phages have been classified according to their structure, type of genetic material (DNA or RNA), host range and infection cycle. Additionally, molecular techniques such as pattern of enzymatic restriction of the DNA, profile of structural proteins, and more recently gene sequencing have been increasingly employed for this purpose. The original studies involving phage aimed the control of pathogenic bacteria in humans and animals, named the phagetherapy. Based on phage estimated number and diversity, it is likely that there are thousands of phages with potential to be applied in several areas. The promising results encouraged the researcher to search for alternative applications such as food control, biofilm dispersion in capillaries and surfaces, specific detection of microorganisms and even as instruments of gene transference for bacterial cells. Thus, a proportional amount of information has been and will still be generated. The gathering of these data in an organized way will support the creation of a viable classification system for efficient implementation of phagotherapy in the prevention and control of pathogens that threaten human and animal health.

[*] Corresponding author: Sérgio Oliveira de Paula. E-mail address: depaula@ufv.br. Tel.: +55 31 3899-2589; fax: +55 31 3899-2549.

1. Introduction

Since the first use of penicillin in the midst of World War II, antibiotics became undoubtedly the greatest tool against bacterial infections. However, the emergence of strains resistant to antibiotics worries. The phagetherapy (PT), or the use of bacteriophages to kill pathogenic bacteria, appears as a potential alternative, but still in its early stages of development.

Bacteriophages are widely distributed, exhibit dramatic expressions in bacterial ecology, and are the most diverse and abundant organisms on the planet, considered by researches the main influence on bacterial evolution. In the late nineteenth century some researchers have suggested the existence of bacteriophages, however their studies did not provide sufficient support for consistent research. In one of them, Hankin (1896) reported that the waters of the Ganges and Jumma rivers have the ability to kill certain types of bacteria, especially *Vibrio cholerae*. He proved that this ability was filterable and sensitive to the boil, which led him to believe that it was a chemical volatile. In 1901, Emmerich and Löw described a substance present in autolysed cultures that had the ability to lyse various other cultures, possessed curative effect in experimental infections, and provided prophylactic immunity to subsequent inoculations. In 1915, Frederick W. Twort published an article describing the growth of some colonies of *Micrococcus*, which had mucoid, aqueous or vitreous appearances, and compared these different morphologies with an acute infectious disease (Twort, 1915). Independently, in 1917, Felix d'Herelle described a "microbe" that act as antagonist of bacteria, lysing them in liquid cultures and forming discrete spots, which he called plaques, on the surface of agar seeded with bacteria. He concluded that these organisms were ultra-viruses that invade bacteria and multiply, and called them bacteriophages (d'Herelle, 1917). Next, some results concerning PT were obtained and excited researches, making phages become a major tool in the control of bacterial infections in humans. However, their uncontrolled use, mainly associated with little knowledge of their basic biology and kinect, made this technology failed in a number of clinical trials in the mid-30's. Furthermore, antibiotics appeared at that time as the promise of the end of the problems with bacterial infections, which contributed to the PT fell into oblivion.

Recently, with the emergence of antibiotic-resistant bacteria, scientists are resuming studies involving PT. As can be seen in Figure 1, there was an increase in the number of publications involving PT in the last decade, as well as the number of patent applications and clinical trials involving the use of phages in the control of pathogenic bacteria, for example with *Klebsiella pneumoniae* (Tsao et al., 2012) and *Pseudomonas aeruginosa* (Sulakvelidze and Pasternack, 2009). However, the development of products based on PT, mainly in Western countries, faces a number of hurdles due to the absence of regulatory mechanisms that specifically cater to the peculiarities of this type of product. To date, the U.S. Food and Drug Administration (FDA) have applied the same rules that regulate the use of antibiotics for phage-derived products, which would significantly increase the cost of the final product and reduce its effectiveness. However this scenario is likely to change. Some phage-based products were recently approved by the Department of Agriculture/Food Safety & Inspection Service (USDA) and the FDA for the control of pathogenic bacteria such as *Salmonella* and *Listeria* in industrial plants and even in ready-to-eat (RTE) food.

Phages generally act over a very restricted host range, allowing the removal of pathogenic bacteria and preservation of the commensal flora, unlike most antibiotics, which indiscriminately acting between both types of microorganisms. However, phage specificity is also a problem, since one must know exactly the strain causing the infection to use the appropriate phage. This can easily be overcome with the use of phage cocktails, which would increase the spectrum of action of the final product, and / or the use of techniques for rapid bacterial identification. Another positive point of using phages is based on their ability to readapt under selective pressure, like the emergence of resistant strains of pathogenic bacteria, what is known as co-evolution cycles. Thus, new adapted phages can be selected at each appearance of a new resistant bacterial strain. Furthermore, cocktails containing different phages, with or without antimicrobials, could be used to achieve a synergistic effect on the pathogenic bacteria and reduce the emergence of resistant strains. On the other hand, the phages pharmacokinetics not favor their use to control infections inside the organism because they could be eliminated by the immune system. Thus scientists are challenged to select phages able to remain for a longer time in the organism or to develop adjuvants which prevent early elimination of viral particles. A second challenge consists in the production in large-scale of viral particles and standardization of solutions to stabilize these particles.

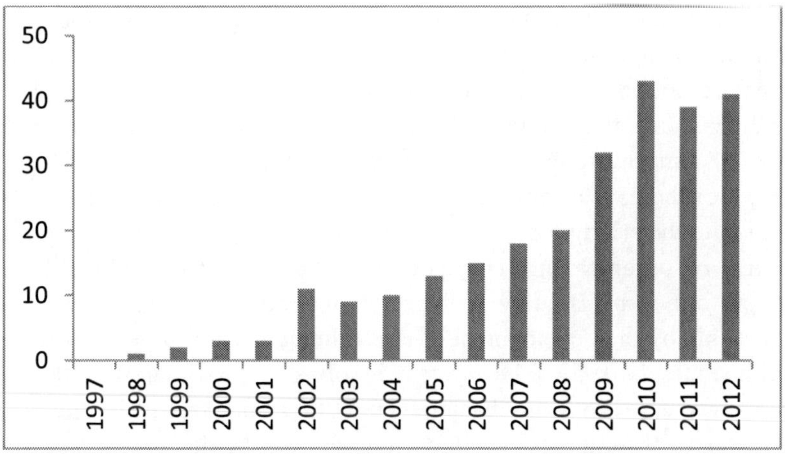

Figure 1. Number of publications in the PubMed Database containing the term "phage therapy" in the title or abstract between 1997 and 2012.

2. Phage Biology

2.1. Phage Structure and Classification

The origin of the caudate viruses during evolution occurred before the separation of the three major domains Eubacteria, Archaea and Eukarya, probably about 3.5 to 3.7 billion years ago. Phages have been reported for Eubacteria and Archaea, but their presence as Eukarya cells-infecting agents remains questionable. Bacteriophages can be caudate, polyhedral, filamentous or pleomorphic (Figure 2) and, except for the caudate, they are not grouped into orders. Caudate viruses are the most abundant bacteriophages, while pleomorphic virions are rare.

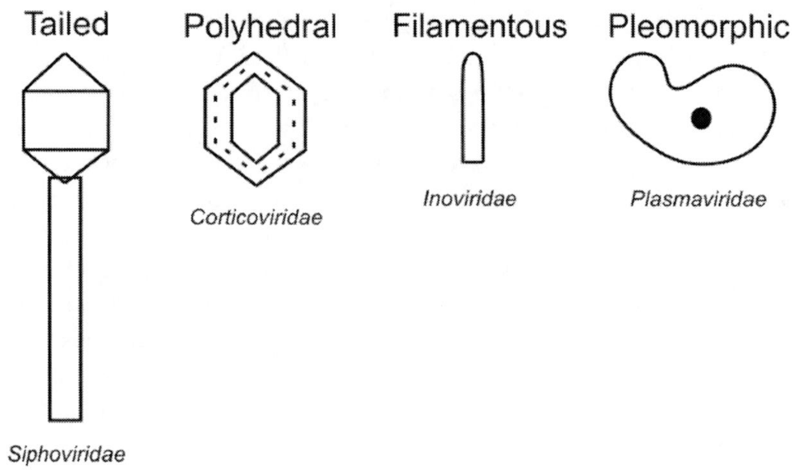

Figure 2. Types of bacteriophages.

A variety of morphologic characteristics can be observed between phage species isolated from natural systems. Typically, a caudate phage has a head (capside) and a tail linked or not by a connector (collar), while cubic, spindle, lemon-shaped, filamentous or pleomorphic shapes are less frequently isolated. Although in great number and types, some structures in phage particles are optional, as the head appendage, collar, and fibers or spikes. The capsid diameter and the genome size can vary by more than an order of magnitude between isolates.

According to current studies, it is estimated that there are about 10^{31} caudate virion particles in the biosphere, requiring somewhere around 10^{24} productive infections per second to maintain this number (Krupovic, M., 2011). Organisms with such extraordinary numbers need some form of organization. The earliest registered attempt of organization of these particles was performed by Bradley in the 60's, and was based on morphological analysis, which were possible by the development of the technique of negative contrasting by Brenner, Horne and colleagues in 1959. This system involved approximately 6300 micrographs of viruses of prokaryote, 6196 from bacterial and 88 from Archaeal. The large difference between the numbers of viral particles in each group may be due to the ease of cultivation of the host. Accordingly to the micrographs, Bradley (1967) defined six different phage morphotypes, which were designated as Myoviridae (contractile tails), Siphoviridae (long non-contractile tail), Podoviridae (short tail), Microviridae (non-caudate, large capsomer), Leviviridae (filamentous) and Inoviridae (non-caudate, small capsomers). Currently, the ICTV (International Committee on Taxonomy of Viruses) admits the existence of one order called *Caudovirales* comprising the caudate viruses, which corresponds to 96% of known, and 7 small families of non-caudate phages, which are not included in any order, some with a single representative.

The order *Caudovirales* is divided into three families: phage with a long flexible tail (*Siphoviridae*), phages with contractile tails (*Myoviridae*) and phages with very short tail (*Podoviridae*). Phages belonging to this order have icosahedral heads or elongated capsids with additional lines formed by subunits starting from central structures. Phages with polyhedral capsids are icosahedral or similar (*Levivirus*) and vesicles lipoproteins can be present inside them (*Tectiviridae*, SH1 and STIV) or as part of the capsid itself (*Corticoviridae*). A wide range of shapes is found in pleomorphic phages, such as fusiform, spherical, bacilliform, and bottle- and drop-like forms. Some virions are enveloped, such as some filamentous phages, which makes them sensitive to chloroform. The phages can possess single or double stranded DNA or RNA as genetic material, which ranges from 3.5 kb (levivirus GA, ssRNA) to 497.5 kb (myovirus G, dsDNA) (Figure 3).

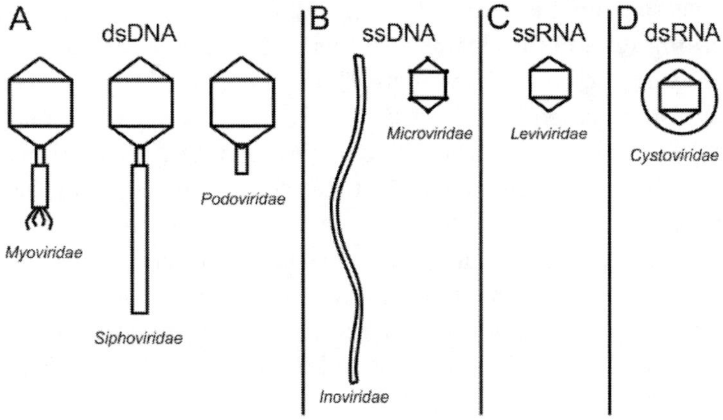

Figure 3. Families of bacteriophages grouped according to their most common nucleic acid composition. A) dsDNA – are represented by tailed phage, which are classified in the order *Caudovirales*, which represents approximately 96% of all phages identified so far; B) ssDNA – mainly represented by the family *Microviridae*, which are the major ssDNA phages found in metagenomics approaches; C) ssRNA – the phage levivirus GA represents this group; D) dsRNA – the *Cystoviridae* family is formed by enveloped virions, as the phage Φ6.

2.2. Infection Cycles

Several steps during the phage life cycle are common to all viruses: adsorption, injection of nucleic acid, expression and replication of the viral genome, particle assembly and virions release.

The adsorption of bacteriophages to the host occurs in two steps. At first, adsorption occurs by reversible interaction of the virion structural proteins (e.g., fimbriae tail proteins) with a component of the cell surface. During the second step, an irreversible binding occurs between these structures. After adsorption, the cell wall is permeabilized by enzymes present in the tail or capsid, allowing the infection of viral genetic material into the cytoplasm of the host cell, while the capsid remains outside. A different mechanism is observed for enveloped phages, as the Φ6 whose envelope merges into the cell membrane, internalizing into the cytoplasm the viral proteic structure surrounding the genetic material. In this environment, the protein structure is disassembled and the viral genetic material is released.

Following the entry of the viral genome in the cytoplasm of the host, two distinct pathways can occur: the viral genome can be integrated into the host chromosome or remain in the cytoplasm as a non-expressed plasmid (lysogenic cycle) or viral genes can be expressed and the vital genome is replicated, followed by the assembly and subsequent release of virions particles (lytic cycle). Cell lysis by caudate phages occurs mainly by a dual system consisting of hydrolases of peptidoglycan (endolysins), which attack the murein layer present in the cell wall, and holins, which damage the cell membrane and allow endolysins to have access to the peptidoglycan. The virions can also be released by budding or extrusion. The phase of the infection in which the phages are internalized in the host cell, which comprehends from adsorption to cell lysis in the lytic cycle, is termed the latency period. The eclipse period is that prior to assembly of the capsid and inclusion of genomes, whereas the number of virions released into the medium by each cell is known as the burst size.

Phages can perform different cycles of infection: lytic, lysogenic, pseudolysogenic and chronic. In the lytic cycle, the host metabolic machinery is diverted to the production of new viral particles, which are released by cell lysis. The lysogenic cycle is characterized by the maintenance of the viral genome in a state of latency (prophage), during which it replicates along with the host genome or as an independent material in the cell, until the lytic cycle is induced. Temperate phages have the ability to migrate between the lysogenic and lytic cycles according to the host characteristics and environmental conditions. In persistent infections (pseudolysogeny or carrier state), only a fraction of the phage population multiplies in the host, while the term carrier is strictly used for bacteria containing plasmid-type prophages. Chronic infections occur when a cell is infected and the viral progeny is constantly released by budding or extrusion.

2.3. Phage Genomics

Until the year 2012 there were about 750 complete sequenced genomes of bacteriophages, comprising several morphotypes and nucleic acid composition combinations. However, the vast majority of these sequences were from caudate viruses with nucleic acid composed of dsDNA. The proportional presence of the caudate families between these phages was about 55, 25 and 20% from the *Siphoviridae*, *Myoviridae* and *Podoviridae* families, respectively. Although the sizes of the genomes varies among the phages, in general the genomic densities in the interior of the capsids are similar, and the size of the capsid varies as a function of the genome size. Too large or too small genomes can influence the stability of the virion, thus influencing viral infectivity. This flexibility favors the viral evolution, since the bacteriophages can acquire or donate genes independently of their functionality. In this context, the theory of modular organization of the phage genome was proposed in the early 80's. and proposed that the evolution of bacteriophages was given by recombination of functional modules, resulting in new viable phages, more complex as they are phylogenetically distanced from their hosts.

Typically, the genes necessary to maintain lysogeny in phages are divergently transcribed and are adjacent to the genes of the lytic cycle. The lysogenic module usually encodes a repressor, which is required for prophage maintenance, and an integrase, which is essential for integration of the viral genome into the bacterial chromosome. The lytic module contains a second repressor, which acts repressing lysogenic module during lytic propagation. Next to the repressor gene are clusters involved in the replication and packaging of the viral genome and in the lysis of the host cell. Studies performed to understand the regulatory control of the lytic-lysogenic change traditionally used the phage λ as a model. It was discovered that this phage has a repressor protein named CI, which interacts to an switch intergenic region preventing the binding of the RNA polymerase to lytic promoters, thereby preventing the establishment of the lytic cycle. This repressor protein also performs an auto-regulated function, since at high concentrations it represses the promoter responsible for its own expression. The CRO proteins exert an opposite effect to that of CI by the attack to the same operators, but with an alternative order of occupation, allowing the occurrence of the lytic cycle. However, additional regulatory elements, including CII, CIII and Q, and complexes interactions between them and host responses to the environment, are involved in establishing both cycles.

A comparison of sequences from different bacteriophage demonstrates an enormous variety in the genomes. Genomes of phages with different hosts have not extensive similarities in nucleotide sequences, suggesting that the host might be a barrier to genetic exchange between bacteriophages. However, phages that infect the same host can present a high genetic variability, suggesting that there are additional barriers and influences in gene exchange. The main feature observed in the phage genome analysis is the mosaicism, with genic segments originated from distinct evolutionary sources. A feasible explanation for this phenomenon is the horizontal gene transfer, which has a key role in the genomic architecture and is easily observable in studies of comparisons of complete genome sequences, Hatfull, (2006) observed great similarity between two closely related Mycobacteriophages (Rosebush and Qyrzula) by comparing their complete genome sequence, When the Rosebush sequence was compared to other mycobacteriophage (Colbert), it was observed low similarity between their nucleotide sequences. However, a single fragment of their sequence had high coverage, with over 94% similarity, possibly due to a recent horizontal transfer of genetic material.

2.4. Ecology

Ecology is the study of the interactions between different organisms, and between the organisms and the environment. One of the relevant factors in ecological analysis is the viral diversity, which is given by the analysis of three parameters: species richness, evenness and diversity. In a system model where species richness is related to the number of species, the equity is the average significance of a single species in terms of abundance, biomass or activity, and the difference is the taxonomic relationship of species in a specific system. Studies of pelagic systems have shown that in deeper regions, less productive systems, viral abundance is low; however, the abundance in coastal regions increases by two orders of magnitude, showing that, in general, the higher the productivity of the system, the greater viral abundance.

Some environmental studies demonstrate that viral exceeds the bacterial abundance by about 10-20 times. Paradoxically, in studies with phage-host systems, when a stable state is established, the host is more abundant than the phage in several orders of magnitude. This paradox could be explained by the existence of non-infective viral particles. Decay data show that about 50% of the viruses exist as infectious particles.

Metagenomics recently caused an explosion in the number of sequences obtained from the most diverse global niches, whether sea, soil, reefs-associated communities, estuaries, among others. This technology can realize the dream of genomic biologists, actually allowing a complete comprehensive analysis of various organisms. The isolation and sequencing of a phage with taxonomic objectives carries some facilities, such as easy cultivation and previous knowledge about the host. The analysis of large numbers of sequences provides relevant information about the phage distribution, diversity and evolution by the Earth. For example, metagenomics confirmed the dominance of caudate phages with genome composed of dsDNA in marine communities (Breitbart, M., 2002, Breitbart, M., 2004). A second study showed the predominance of certain viral types in specific regions, such as the prevalence of phage with ssDNA in the Sargasso sea, and the prevalence of prophages in the Arctic (Angly, FE, 2006). Other findings in these studies were the presence of genes of eukaryotic viruses, as genes of phosphate-sensibility, response to stress and photosynthesis, all present in their respective hosts (Comeau, AM, 2008).

Due to predation exerted by phages on microbial communities, these viruses exert a strong influence on the carbon and nutrients flows and food dynamics, showing that bacterial competence is not only related to their adaptation to available resources, but also to biotic agents present in the environment. To combat phage activity, bacteria have developed defense systems, among which we can mention: i) preventing adsorption, hindering the interaction of viral proteins to the cell receptors by changing the receptor expression, by masking receptors with other molecules, by the production of exopolysaccharides (EPS), etc.; ii) blocking DNA injection through changes in the membrane permeability; iii) restriction-modification strategies, using cellular methylases which modify the cellular genome making it inaccessible to endonuclases, but allowing the enzymes to act on the foreign DNA; and iv) abortive infections, in which the cell interferes in one of the steps eh the lytic cycle, acting on replication, transcription, translation or assembly steps. Following this strategy the cell is dead and no infectious viral particles are released, preventing neighboring cells to be infected.

Several studies have been conducted on the use of bacteriophages as alternative tools to the conventional drugs, mainly due to the emergence of resistant bacterial strains. Some studies regarding the phages biology and proteomics difficulties hamper the regulatory approval process. However a renewed interest in PT has been observed in the scientific literature. Part of this interest stems from the appreciation of the extraordinary abundance of these particles in the biosphere, allied to a series of experiments conducted with phage lysins, to the large number of experiments with animals using viable phages as antibacterial agents and to historical revisions of the Soviet experience with PT that are currently being recovered.

3. Applications

The world is in the midst of a crisis in relation to disease control. After decades of indiscriminate use of antibiotics and other antimicrobials, multiple resistant bacterial strains have emerged and led to epidemics that are becoming increasingly difficult to control with conventional antimicrobials. Thus, the search for alternative, convenient, cheap and safe methods has been the focus of a growing number of research centers and funding agencies worldwide.

Most of the scientific research carried out in the second half of the last century involving the use of bacteriophages to control infections was restricted to Eastern Europe, where PT is a reality. This difference was exacerbated due to the political problems of the Cold War and the linguistic and cultural differences of the Western World. Among the scientific studies and clinical tests conducted in the twentieth century involving PT, a large number is not available in international journals of English language. Part of the data generated in these studies was lost, while the rest has been published in books on PT.

Phages are natural predators of bacteria and are commonly found in the environment in which its host inhabits. It is believed that phages are responsible for the elimination of 4 to 50% of the total bacteria produced every day, contributing significantly to the geochemical cycles and as a reservoir of genetic diversity, serving as a natural, highly specific, non-toxic and accessible technique for the control of various micro-organisms. Its use as a biological control agent has increasingly become a viable option for reducing the bacterial population in various environments. Among the many advantages of PT are its efficiency against pathogenic bacteria resistant to antimicrobials; the viral specificity, ensuring the integrity of commensal bacteria; the ability of these particles to quickly respond to the formation of phage-resistant bacteria, in coevolution cycles; the low cost of production of phage-based products; the safety to eukaryotic cells, where side effects of its use are uncommon; their widespread presence in large numbers and diversity in nature; and its easy cultivation and manipulation.

Factors such as the multiplicity of infection (MOI), the number of doses and the route of administration of these doses could be critical to the efficiency of bacterial inactivation. For example, it has been demonstrated that the reduction in the number of bacteria is proportional to the MOI. In contrast, unlike antibiotics, a single dose of phage could be sufficient to kill 100% of the bacterial population. The route of application has a direct implication on the efficiency of PT. Historically, the main route of administration is oral, which reduces the effects of contaminants of the phage preparations, such as endotoxins. However, in this case the viral particles could be subjected to conditions which may result in its inactivation.

The pharmacokinetics of phage-derived products is poorly understood and more complex than chemotherapy. This is because to its dependence of the characteristics of the pathogen, the individual (animal or human), and the multiplicative ability of the phage or phage cocktail. It is therefore very important to ensure that proper dosage of infective viral particles reach the site of infection.

Another point to be considered is the partial or complete elimination of the phage by the innate immune system of the treated individual, since a decrease in phage titre in the blood of mice injected with phages are generally attributed to elimination by the natural defenses of the organism. Techniques as serial passage have been used to isolate phage mutants with a greater capacity to remain in the circulatory system of the mouse. These 'long-circulating' phages were shown to be more effective therapeutically. The adaptive immune response of the individual may lead to the phage neutralization or adverse reactions such as anaphylactic shock, although, to date, there are no reported cases of anaphylactic shock caused by PT. If a particular phage strain is repeatedly used in a single organism, the adaptive immune system would be stimulated and result in the production of antibodies in a few days by clonal expansion of lymphocytes. Depending on the immunogenic portions of the phage proteins, these antibodies could block infection of bacterial cells via phage neutralization. If a new injection of phage occurs, a secondary response will happen faster and produce a greater number of antibodies. Thus, the pharmacokinetics of PT varies with the immunogenicity of the phage proteins and with the number of doses administered in each patient. Beyond neutralization, antibodies against the phage structural proteins could also cause the acceleration of phage clearance in the body. In fact, patients with severe combined immune deficiency display a prolonged period for the phage clearance.

3.1. Using Phages to Control Bacterial Infections

In the following topics we summarize the main studies performed in order to enable the use of phages for the prevention and treatment of bacterial infections.

3.1.1. Using Phagetherapy to Treat Skin Infections and Burn Patients

The PT has been successfully used to prevent rejection of skin grafts in animals, to treat skin infections and to prevent and treat infections in patients who have suffered burns. The bacterium *P. aeruginosa* is the major pathogen involved in skin infections. This bacterium is also the causative agent of other diseases such as pneumonia, bacteraemia, meningitis and urinary tract infections in individuals with suppressed immune systems. Moreover, bacterial strains of *P. aeruginosa* are commonly found as multiresistant cells, frequently isolated from biofilms on hospital equipment and utensils.

The high mortality rates among patients with severe burns have led researchers to reevaluate the conventional methods used in their treatment. The main cause of death of burn patients within the first 2 days occurs due to infections by bacteria *P. aeruginosa*, *Staphylococcus aureus*, *Escherichia coli* and *Streptococcus*, which are currently considered serious problems in hospitals due to the presence of various strains resistant to traditional antibiotics. Accordingly, topical application of phages-based products, or its use as active particles adsorbed onto polymers and artificial skin grafts (PhageBioderm), have been developed in specialized research centers and have shown promising results.

The study performed by Lazareva et al. (2001), for example, showed that the use of phages to treat infections in wounds from burns promoted the disinfection of wounds and normalization of the body temperature and the levels of immune system cells, with consequent reduction in mortality. In 2009, Merabishvili and collaborators obtained promising results using a phage cocktail called BFC-1, which consisted of one *P. aeruginosa*-specific phage and one S. aureus-specific phage, what led this group to further studies of these agents, including its effectiveness against antibiotic-resistant strains, infection parameters, stability, genomic composition, studies of routes of application, time of residence in the body, etc.., attempting the feasibility of its use in Western medicine.

3.1.2. Using Phagetherapy to Treat Urinary Tract Infections

Urinary tract infections (UTI) are usually caused by *E. coli*, but other Enterobacteriaceae can also cause this type of infection. The colonization of the urinary tract confers to the pathogen a protective environment, so it could increase up to 100 times its tolerance to antibiotics. In this context, PT appears as an interesting and promising for the control of this type of infection, since phages are specific to a particular strain or group of bacterial strains. This feature ensures only the elimination of the pathogen and the maintenance of commensal flora, what is particularly important in UTIs.

In vitro and *in vivo* studies demonstrated the potential of PT to the treatment of UTIs, although controlled clinical studies are still scarce, particularly in Western countries. For example, the phage T1, a T4-like phage, proved to be able to cause a 45% reduction of different *E. coli* strains (isolated from UTI patients) attached to epithelial cell monolayers after two hours of treatment (Sillankorva et al., 2011). Another T4-like phage, KEP10, was effective in controlling enteropathogenic *E. coli* (EPEC) in mice model when inoculated intraperitoneally (Nishikawa et al., 2008). This phage, when inoculated by this route of administration, was able to prevent the death of mice experimentally infected with the pathogen. Interestingly, KEP10 was present at high levels in different organs for at least 24 h. This situation is fairly different from the case of phage lambda, which was rapidly eliminated by the organism. Thus, one of the points to be considered during the selection of an appropriate phage for use in PT is its low immunogenicity and a favorable pharmacokinetic, as previously described.

In Western medicine there are very few clinical studies of PT against UTIs in humans. In Russia, the sensitivity to phages of pathogenic strains isolated from 46 different UTIs patients was analyzed in order to optimize the formulation of preparations suited to each patient. The clinical efficiency of this study reached 90% when the oral and local routes of administration were used (Perepanova et al., 1995).

Thus, although still scarce, the existing studies are able to show us that PT is a potential candidate for the control UTIs. Currently, researchers' biggest challenge is to construct a set (phage bank) of practical therapeutic phages able to infect and kill pathogens associated to this type of infection, and proved good agents for PT of UTIs, which will only be accomplished by well-controlled *in vivo* clinical studies.

3.1.3. Using PT to Control Respiratory Tract Infections

Respiratory infections are a major medical concern due to the high degree of mortality among affected patients and also due to the recent emergence of a large number of strains isolated from sick patients that are multi-resistant to the conventional antimicrobial agents. Among the most worrisome diseases are chronic lung infections in patients with cystic fibrosis (CF), for which alternative treatments have been evaluated due to the high failure rate in conventional treatments.

Patients with CF are usually affected by lung infections caused by *P. aeruginosa* which consists of one of the main problems in antibiotic therapy due to their ability to form biofilms and acquire resistance to conventional antimicrobials. Several phages have been isolated in attempt to eliminate this pathogen from lung, and some clinical trials have been performed. For example, phages NH4 and MR299-2, isolated by the group headed by Alemayehu in 2012, were active against nine clinical strains of *P. aeruginosa*, and a mixture containing these phages was effective in eliminating the pathogen in murine lungs and also when this bacteria was forming biofilms in human lung cell lines, representing an important demonstration of efficacy in validated infection models.

A second group of researchers led by Debarbieux, of the Pasteur Institute, has investigated the action of phages in the prevention and cure of pneumonia caused by a strain of *P. aearuginosa* isolated from patients with CF. Their experiments showed that instillation of a phage-containing solution at 24 h before the inoculation of mice with the bacteria prevented the occurrence of the lethal form of the disease, with a survival rate of 100%, while all animals which were not instilled with the phage solution died. In the same way, they observed a 75 and 25% survival rate when phages were instilled 4 and 6 h after challenge with bacteria, respectively. In another study performed by the same group, a total of 95% survival was observed after intranasally inoculation of phages at 2 h after infection with the bacterium, and a four-day preventive treatment (one single dose) resulted in a 100% survival.

Despite the successes achieved in these and other treatments, the route of administration of the doses remains one of the main issues in the optimization of treatment for respiratory tract infection because bacterial infections in the lung may reside in macrophages or as colonies in the vasculature protected by structures such as mucus layers or biofilms. Among the routes already tested, inhalation / nebulization seem to be efficient in deliver phages to the lungs. This strategy of administration is also promising because many phages can be stabilized in the solid state for years, which has led to the research of dry powder inhalers containing phages incorporated into microparticles.

3.1.4. Using Phagetherapy to Control Pathogens in Animals

Bacterial infections in animals cause priceless damages to agriculture, even in countries where sanitary control is considered satisfactory. Several antimicrobials used for industrial management, some of which are freely sold as growth promoters, are known to cross-react to those used in human medicine, making their use increasingly discouraged. Furthermore, the presence of antibiotic-resistant bacteria has been demonstrated in a number of animal products, especially dairy and beef, widely consumed by humans. Because this area has prime economic importance to the commercial balance of a large number of countries, the search for safe and inexpensive alternative antimicrobials is a priority, what makes this issue a major funding source of research in PT.

Control of Salmonellosis in Poultry

Despite the evolution of industrial poultry, salmonellosis is still a limiting factor in the creation of birds, occupying a prominent position due to economic losses ranging from chicken production to final consumption of poultry byproducts. Salmonellosis is the term used for acute or chronic diseases caused by bacteria of the genus Salmonella. The birds, once contaminated, become intestinal carriers and eliminate the pathogen in the faeces for longer periods of time, reaching the slaughterhouse, compromising the entire production line. The epidemiological and economic implications of foodborne illness in humans can achieve great magnitude, where the increasing incidence is associated with ingestion of origin-contaminated and / or improperly prepared poultry products, since the birds are considered *Salmonella* spp reservoirs.

Different serotypes of *Salmonella* are able to cause salmonellosis, which makes the use of phages to control it even more challenging. There are over 2500 serovars of *Salmonella enterica*, of which 13 can be commonly isolated from poultry in the U.S., including Enteritidis and Typhimurium, which are also the serovars more frequently isolated from human sources in the U.S.. There is not known any phage that is able to infect and kill all *Salmonella* serovars. More likely, a particular *Salmonella*-specific phage will lyse only a small part of the spectrum of *Salmonella* serovars. A higher degree of host specificity requires the use of phage mixtures for prophylaxis of bacterial infections in most instances, as made by Chighladze et al. (2001), which created a phage cocktail that was able to lyse 232 of 245 *Salmonella* isolates representing 21 serovars and reduced Salmonella to undetectable levels after 48 h in artificially contaminated surfaces. Although several *in vitro* studies have obtained the removal of 100% of the initial amount of *Salmonella* using specific phages, different, but not less exciting, efficiencies were observed during *in vivo* tests with *Salmonella*-specific phages. These studies showed reductions in the counts of colony forming units ranging from 2 to 4 $\log_{10}$ units, indicating that PT can be applied to salmonellosis control in birds even considering the differences between the *in vivo* and *in vitro* results.

As occurs for studies on the use of PT to control infections in humans, *in vitro* studies are revealing the great potential of PT for the control of salmonellosis in birds, although *in vivo* studies show that there are still many questions to be answered before PT becomes a reality. Among the factors that may explain these differences is the lack of knowledge of phage pharmacokinetics, with the possible loss in the number of viral particles during passage through the tract of the animal. Strategies of microencapsulation to provide protection for the particles in the gastrointestinal tract have been attempted. A promising association for using phages in the poultry industry is their association with products based on competitive exclusion such as probiotics. The combined application of these products presents a synergistic action in reducing the count of colony forming units of *Salmonella*. The PT, being a biological mechanism, provides quality for the food, whereas phage particles not leave residues in meat, eggs and also the environment, besides being economically viable when compared to the cost of traditional antimicrobials. These factors, associated with the inherent biosecurity on the using phages, enable the label of "promising" for the PT in the control of avian salmonellosis.

Control of Bovine Mastitis

Despite several control programs, bovine mastitis represents the most economically important disease of dairy cattle and remains a central problem on dairy farms. It causes reduced milk production, increased treatment costs, extra labour requirements and an increased rate of culling. Mastitis is most often caused by infectious agents, such as *Staphylococcus aureus*, *Streptococcus agalactiae*, *Streptococcus dysgalactiae*, *Streptococcus uberis*, *E. coli*, *Citrobacter freundii*, *Enterobacter* sp., *Klebsiella* sp., *Pseudomonas aeruginosa* and *Arcanobacterium pyogenes*. These pathogens invade the mammary gland, especially those glands predisposed to infection due to trauma, and produce toxins that injure the milk-secreting tissues and ducts throughout the gland.

Studies involving the isolation and identification of bacteria from milk obtained from animals with mastitis report the prevalence of coagulase-positive *Staphylococcus*, which frequently exhibit a broad spectrum of antibiotic resistance. Many studies have emphasized the lack of genetic diversity among multiple-drug-resistant bacteria, probably due to clonal selection, what means that phages are ideal agents for eliminating these organisms. This has led researchers to create libraries of phages with different specificities and efficiencies to test them in combating this disease.

A major concern of medical and veterinarians is the increased number of cases of isolation of methicillin-resistant strains of *S. aureus* (MRSA) from the most diverse environments. This bacterium is widely associated with cases of bovine mastitis, and the use of phages is being tested to combat this ill-considered of medical emergency. Kwiatek, M (2012) reported the isolation of a myophage from cows with mastitis, the MSA6, which showed lytic activity against a broad host range of staphylococci both associated with mastitis and related to human infections. This phage exhibited rapid adsorption and a short latent period considered promising characteristics for use of phage in PT.

3.2. Biocontrol of Foodborne Pathogens

Foodborne diseases are responsible for high levels of morbidity and mortality in the general population, but particularly for at-risk groups, such as infants, children, the elderly and the immunocompromised individuals. The Center of Disease Control (CDC) estimates that 9.4 million illnesses, 55,961 hospitalizations, and 1,351 deaths occurs each year, only considering the majors known foodborne pathogens in United States. However, these numbers are much higher if we consider unknown agents and non-reported cases. Recently, bacteriophages have been proposed, and used, as biocontrol agents in human foods due to their ability to promote disinfection of contaminated food. In this case, the main targets are *Listeria monocytogenes*, *Salmonella*, *E. coli* O157: H7 and *Campylobacter* due to their relevance in human health. Although there is still difficulties in the registration of phage-based products in the Western world, the FDA, in 2006, allowed the use of a cocktail of 6 different phages effective against 170 strains of *Listeria* in meat products, the INTRALYTIX's. This was the first phage-based product approved by the FDA to be applied to ready-to-eat foods, which represents a clear and formal recognition of the phages as realistic antibacterial agents in foods. Similarly, ListexTM (MICREOS Food Safety, Netherlands) and OMNILYTICS (Salt Lake City, UT), a hide-washing solution containing phage active against *E. coli* O157: H7, have obtained their approval from the FDA and the USDA, respectively. So the phage-based products begin to gain the confidence of Western governmental control agencies.

Unlike what happened in the past with antibiotics, phages alone are not to be a substitute for antibacterial agents as an independent technology. Although using pure phages has shown high efficiency in controlling bacterial contaminants in food, its association with other control processes has shown synergistic results. A combination strategy that has been adopted to control *Listeria* is the association of specific phages with nisin, a peptide produced by lactococcal strains with antimicrobial properties. This is known as barrier technology, which employs different and simultaneous technologies to control food spoilage, and has been widely used in the food industry, in order to reduce the negative effects of each treatment on the physico-chemical and sensorial characteristics of the food. Another important factor is the reduction in the selection of resistant bacterial strains due to high selective pressure generated by the combination of methods. Thus, the phages are logical candidates for the control of undesired bacteria in foods, especially when used in combination with other methods of control, and can make a valuable contribution to food safety and public health.

3.3. Using Phagetherapy to Control Biofilms

A great number of microorganisms is able to colonize indwelling medical devices to form a biofilm. The biofilm is a structure composed of a consortium of microorganisms colonizing a surface where the different species are integrated and surrounded by a complex structure formed by extracellular polymeric substances (EPS) produced by them. The various molecules in the matrix include cell surface proteins, pili, DNA, RNA, lipids, and especially polysaccharides. The exact composition of the biofilm depends on several factors such as species of microrganisms that are present, types of nutrients available and the material from which the adherent surface is composed. The more common microorganisms found in biofilms from indwelling medical devices are coagulase negative staphylococci, *S. aureus*, *Enterococcus faecalis*, *Enterococcus faecium*, *E. coli*, *Proteus mirabilis*, *Klebsiella pneumoniae*, *Viridans* streptococci, *P. aeruginosa* and *Acinetobacter* spp.. It is common that bacteria in the biofilm are catalysts for chemical reactions that cause corrosion of equipment, mainly by the release of sulfur compounds. The presence of biofilms may also cause clogging of capillaries, reducing of the heat flow over the surfaces and releasing of adhered cells, resulting in bloodstream or urinary tract infections.

Commonly, biofilm organisms are tolerant to antimicrobial agents, disinfectants and biocides, e the treatment of device-associated infections with systemic antimicrobial agents is usually ineffective, making high concentrations of antimicrobials for prolonged treatment durations normally required. In this context, the use of phage for the treatment of device-associated contaminations could prevent biofilm formation on these devices, reduce the use of antimicrobial agents, and limit the spread of antimicrobial-resistant organisms.

Several phages have been tested to their ability to inhibit formation of these structures or destabilize biofilms. However, infection of cells in a biofilm by phage is highly conditioned to biofilm chemical composition and environmental factors such as temperature, stage of development, physico-chemical conditions of the medium and phage concentration. Studies have demonstrated the effectiveness of these particles under conditions of nutrient limitation, as occurs in a hospital environments, which reduces the rate of bacterial protein synthesis and consequently limits viral multiplication. For example, phage T4 was effective against *E. coli* biofilms in glucose-limited chemostat (Corbin et al., 2001) and higher nutrient levels result in increased cell growth rates. Other studies have demonstrated that there is no need for prolonged exposure times to phages efficiently reduce the bacterial population in biofilms, and that biofilm age did not significantly decrease the susceptibility of *P. aeruginosa* biofilms to phage F116 (Hanlon et al., 2001), unlike what occurs with antimicrobial whose minimum concentration must be increased according to the biofilm age. Regarding the diffusion limitations that occur in the application of antimicrobials, several studies have reported the ability of phages to spread through the polysaccharide matrix of the biofilm by means of various mechanisms such as moving through biofilm water channels and cell clusters to the biofilm interior of its host bacterium or by the production of polysaccharide depolymerases with potential to degrade the biofilm EPS matrix. These studies indicate that some of the most important obstacles to biofilm control by antimicrobial agents might be overcome by phage.

The most effective treatments are concentrated in prevent biofilm formation rather than control existing biofilms. To use the phage in preventing biofilm formation on medical indwelling catheters or other devices, the ideal scenario would be that in which virus particles were adhered on the surface thereof. For this, some important point to be verify are the capacity of the coating matrix to adsorb the phage, infectivity and stability of the incorporated phage, the efficacy of the incorporated phage in the presence of serum proteins and tolerance of the coating by the patient.

Genetic engineering has been a great aid in optimizing the techniques of biofilm inactivation by increasing the efficiency of phage via the changing of its genome, either to insert a gene encoding a depolymerase able to degrade the biofilm polysaccharide matrix, either by withdrawal of lysogeny genes, which may decrease the phage effectiveness. The persistence of device-associated infections drives the quest for new treatment strategies that target biofilm-associated organisms.

3.4. Using Phages to Detect their Hosts

The use of specific fluorescence-labeled phages (Fluorescent Bacteriophage Assay – FBA), phages containing the gene for luciferase (Luciferase Reporter Phage – LRP), or phages that encode the green fluorescent protein (GFP) have been considered an economically viable, sensitive, practical, and fast alternative for the detection of microorganisms in a number of environments. For example, a specific *L. monocytogenes* phage containing the *lux* gene was used to detect this contaminant in meat and showed great sensitivity and practicality, even in the presence of large numbers of natural competitors (Loessner et al., 1996).

A successful example is the detection of *Mycobacterium tuberculosis* using specific phages. For decades phages were tested for their ability in the identification, diagnosis and susceptibility testing of tuberculosis. From these studies were created rapid, simple and relatively inexpensive tests to detect this microorganism (FASTPlaqueTB and PhageTekMB, Biotec Laboratories, UK). These tests rely on the formation of lysis plaques on crops and / or on the use of phages genetically modified to expressing chromogenic or fluorescent proteins following the host infection. Additional details about these and other methods used for detection of bacteria using bacteriophages are described in the work of Hagens & Loessner (2007) and Schofield et al. (2012).

The disadvantages of using this technology include questions regarding the safety and feasibility in using active phages and the need for viable pathogens to the success of the tests. These hurdles have led researchers to devise alternative models of detection using viral proteins responsible for host recognition. These proteins are highly specific and have been cloned and expressed in heterologous systems for this purpose, which will be treated in the next section.

3.5. Using Phage Recombinant Proteins

Detection of Pathogens Using Recombinant Viral Proteins

As shown above, genetically modified bacteriophages carrying reporter genes can be used to detect specific bacterial strains. A refinement of this technique is the use of fiber proteins of the phage tail, which are highly specific to a particular type or group of pathogens. These proteins are normally produced in heterologous recombinant systems connected to a chromatogenic substance so that color intensity can be measured and is proportional to the interactions between the labeled recombinant protein and the bacterial component, a technology similar to the ELISA test (enzyme-linked immunosorbent assay) which uses labeled antibodies for the detection of specific proteins. However, phage proteins can generate more sensible and specific tests than those using antibodies. The process of detection by phage proteins is carried out with only a specific protein and not the infectious particle, and is therefore considered safer for use in industrial processes. A product based in this technology was actually patented and is commercialized as VIDAS UP® by BioMérieux Industry, France. Currently, there are available kits for the detection of *Salmonella*, *E. coli* O157 and *Listeria* in samples of food, feed, environmental samples and soil.

Besides these, there are still endolysins-based tests. Endolysins are enzymes produced at the end of the lytic cycle of dsDNA bacteriophages and are capable of hydrolyzing peptidoglycans from the bacterial cell wall, thus facilitating cell lysis and therefore the release of viral progeny. The PlyG is a endolysin of the phage Gamma and is able to specifically interact and kill germinating spores of *Bacillus anthracis*. The interaction between this protein and bacterial sprouting cell promotes a rapid increase in the release of intracellular ATP, which can be detected in the presence of luciferin / luciferase, readily producing light in proportional intensity to the amount of released ATP.

Using Endolysins to Control Pathogens

Recombinant or purified endolysins are able to destroy the bacterial cell wall when externally applied, resulting in rapid lysis of the bacterial cell. This kind of direct application is only effective against Gram-positive bacteria such as *S. pneumoniae* and *B. anthracis* since the outer membrane of Gram-negative prevents contact of this enzyme to the cell wall. However, once each endolysin have one or more specific hydrolytic activities (amidases, peptidases and glycosidases), and the components of the cell walls may vary between different bacterial species, it is necessary to perform tests to identify its spectrum of action. In addition, due to these different activities, the efficacy of lysins may be further enhanced synergistically by using a combination of two enzymes of different enzymatic specificities.

The process of external application of purified endolysins to control pathogenic bacteria *in vivo* has been studied since the last decade. Some potential problems associated with their use, such as its immunogenicity and release of pro-inflammatory components during bacterial lysis, in fact do not seem to render them unfit for use. In addition, with their independent functional domains, makes them ideal for domain-swapping studies in which bacterial specificities and catalytic activities are improved or adapted for use with other pathogens. This also includes genetic structural engineering strategies to increase their specificity. An example is the association of endolysins with specific recognition proteins or pore-forming proteins for the elimination of specific pathogens, including gram-negative bacteria such as uropathogenic *E. coli* (UPEC). The following paragraphs describe two examples of successful use of purified lysins to control infections *in vivo*:

Selective elimination of S. pneumoniae (Loeffler et al., 2001): the Dp-1 Pal amidase (Pal), a bacteriophage endolysin expressed in heterologous prokaryotic system, was purified and is already successfully used in *in vitro* and *in vivo* tests of bacterial inactivation. Fifteen strains of *S. pneumoniae*, two "capsule-deficient" mutants of *Pneumococcus* (R36A, Lyt 4-4), beyond three penicillin-resistant strains and eight species of streptococcus from the human commensal flora (*S. gordonii, S. mitis, S. mutans, S. oralis, S. salivarius, S. intermedius, S. crista, and S. parasanguis*) were submitted to the purified enzyme for 30 seconds. The result showed that both the 15 strains of *S. pneumoniae*, as the penicillin-resistant mutants and the "capsule-deficient" were equally susceptible to Pal, reducing by about 4.0 $\log_{10}$ the number of colony forming units (CFU) of these bacteria, while two species of commensal bacteria, *S. mitis* e *S. oralis*, were almost completely eliminated. On the other hand, the other commensal species were not susceptible to this enzyme, indicating its high specificity. Tests *in vivo* in mice intranasally infected with *S. pneumoniae* and treated with Pal showed that treatment with Pal reduced the number of *S. pneumoniae* to undetectable levels in all animals, while control treatment detected on average 3.0 $\log_{10}$ CFU/10 μL.

Treatment of B. anthracis infections (Shuch et al., 2002): the purified PlyG endolysin was used in *in vitro* and *in vivo* tests against different species of *Bacillus*. Fourteen strains of *B. anthracis*, ten strains of *B. cereus* and four of *B. thuringiensis* were treated with the enzyme, which was specifically effective against all *B. anthracis* and only one of *B. cereus* (RSVF1), which is related to *B. anthracis*. Thus, endolysins are potential antibacterial agents that rapidly and selectively kill pathogenic bacteria. Thus, considering the estimated total population of dsDNA phages (more than 10^{30}), the technology of employing endolysins to control bacterial pathogens has an enormous potential that remains to be explored. Bacteriophages and their corresponding lytic enzymes activities represent an enormous untapped pool of agents with which to control human pathogens.

Viral Integrases

Viral integrases have the capability of integrating plasmids and other DNA fragments in chromosomal sequences, preventing plasmid clearance after successive cell divisions. This feature gives to these enzymes a great potential for use in gene therapy for the treatment of various diseases of genetic origin. Therefore, integrases have been used for integrating plasmids containing genes of interest to mammalian cells, or for integrating complementary sequences of specific genes for the repression of the expression of a particular protein by the cellular metabolic machinery (iRNA). As an example, the integrase of the phage ΦC31 has been extensively studied as an integration system used in adenoviral vectors, which has been showed an attractive alternative for the engineering of cells, because it allows the efficient site-specific integration of large DNA fragments in the most diverse cell types.

Conclusion and Perspectives

Bacteriophages are the most abundant organisms on Earth and are structurally composed of a capsid, attached or not to a tail, with optional structures present in many of them. This proteic structure is involving their genetic material, whose composition is highly variable. Due to this diversity, the phages were traditionally classified based on their morphology and, more recently, on sequence similarities and genomic organization. Currently they are classified in one order (*Caudovirales*) which comprises three families, and in other 7 families that are not grouped in any order. The phages can infect cells displaying the lytic, lysogenic, pseudolysogenic or chronic cycles, which may be influenced by physiological conditions of the host and its environment. The study of the molecular factors that influence the phage decision on the life cycle is of extreme importance, since the horizontal gene transfer and the presence of host genes in the viral genome are common features in different phages. The entry of phages in the Metagenomics Era can generate relevant information about their genomic organization and thus answer questions relating to the mechanisms of bacterial resistance, as well as the interactions between the phages and many other organisms.

As shown in this chapter, phages can be applied in various types of organisms, by different routes of administration. Since often a single bacterial species is the causative agent of different diseases, as happens in *P. aeruginosa*-associated diseases, an unique phage can be used as agent of PT for different diseases, changing only the dose and route of administration so that its quantity and activity is maximized at the site of infection. Furthermore, recombinant phage proteins have been successfully used as specific and sensible agents of diagnostic and control of bacterial pathogens.

Among the many advantages of using these agents to control pathogens, we can cite phage high specificity, protecting the patient commensal flora; phage ability to exponentially growth inside the host, what optimizes their pharmacokinectics inside the individual; their effectiveness against multi-drug-resistant pathogens; their synergistic effect when administered in association with others antimicrobials; the facility to manipulate virion particles; phage safety, since they are non-toxic to animals and human cells. Moreover, many of these particles are able to synthesize proteins which degrade the biofilm matrix, increasing phage effectiveness. Those proteins can also be used as purified proteins to be applied as non-infective antimicrobial agent. On the other hand, the phage specificity is also a hurdle in PT because it requires the pathogen identification at species-level before the treatment. Furthermore, the original pharmacokinetics of the isolated phages might not be ideal for their use in PT, and immune response could hinder their use. To overcome these problems, an appropriate and careful selection of less immunogenic particles and virions with satisfactory pharmacokinetic can be achieved after multiple passages *in vitro*. Bacterial resistance is also commonly cited as one of the problems associated with the use of phages. However, phages develop in coevolution cycles relative to its host, continuously giving rise to particles adapted to the resistance of the bacterial cell. There is also a apprehension in the use of phages because they can act as genes carriers between bacteria, although the selection and genetic engineering of phages has supplanted these questions, as well as the use of phage recombinant non infective proteins.

In 2012, the World Health Organization declared that the antibiotics era began to reach its end, reporting its apprehension that trivial infection agents become major public health disasters. Among the alternatives, PT is still not seen, at least by Western countries, as a real strategy. Thus, the main question that remains to be answer is: Will phages be no longer a potential pathogen-specific killer to become a real strategy applied to benefit the human life? Brüssow, a leading researcher in the area of PT in the world, recently said that for the PT becomes a reality in countries with stricter rules of control, the researches need to focus their efforts to understand the phage biology and behavior in their natural environment and in the human body, as well as the parameters that influence their interactions with the host, to thereby reduce the large discrepancies observed in the results of *in vitro* and *in vivo* studies.

Acknowledgments

This study was supported by grants from the Fundação de Amparo à Pesquisa do Estado de Minas Gerais (FAPEMIG), Coordenação de Aperfeiçoamento de Pessoal de Nível Superior (CAPES) and Conselho Nacional de Desenvolvimento Científico e Tecnológico (CNPq). The funders had no role in the study design, data collection, analysis, decision to publish, or preparation of this manuscript.

References

Ackermann, H. W., M. S. Dubow (1987). Viruses of Prokaryotes. In: *General Properties of Bacteriophages*, Vol. *I*. CRC Press, Boca Raton. 202 p.

Ackermann, H. W. (1996). "Frequency of morphological phage descriptions in 1995." *Archives of virology 141*(2): 209-218.

Ackermann, H. W. (1998). "Tailed bacteriophages: the order caudovirales." *Advances in virus research 51*: 135-201.

Ackermann, H. W. (2001). "Frequency of morphological phage descriptions in the year 2000. Brief review." *Archives of virology 146*(5): 843-857.

Ackermann, H. W. (2007). "5500 Phages examined in the electron microscope." *Archives of virology 152*(2): 227-243.

Ackermann, H. W. and D. Prangishvili (2012). "Prokaryote viruses studied by electron microscopy." *Archives of virology 157*(10): 1843-1849.

Alemayehu, D., P. G. Casey, et al. (2012). "Bacteriophages phiMR299-2 and phiNH-4 can eliminate Pseudomonas aeruginosa in the murine lung and on cystic fibrosis lung airway cells." *MBio 3*(2): e00029-00012.

Angly, F. E., B. Felts, et al. (2006). "The marine viromes of four oceanic regions." *PLoS biology 4*(11): e368.

Atterbury, R. J., M. A. Van Bergen, et al. (2007). "Bacteriophage therapy to reduce salmonella colonization of broiler chickens." *Appl Environ Microbiol 73*(14): 4543-4549.

Biswas, B., S. Adhya, et al. (2002). "Bacteriophage therapy rescues mice bacteremic from a clinical isolate of vancomycin-resistant Enterococcus faecium." *Infection and immunity 70*(1): 204-210.

Borie, C., M. L. Sanchez, et al. (2009). "Aerosol spray treatment with bacteriophages and competitive exclusion reduces Salmonella enteritidis infection in chickens." *Avian Dis 53*(2): 250-254.

Bratbak, G., M. Heldal, et al. (1996). "Dynamics of virus abundance in coastal seawater." *FEMS Microbiology Ecology 19*(4): 263-269.

Breitbart, M., B. Felts, et al. (2004). "Diversity and population structure of a near-shore marine-sediment viral community." *Proceedings. Biological sciences / The Royal Society 271*(1539): 565-574.

Breitbart, M., P. Salamon, et al. (2002). "Genomic analysis of uncultured marine viral communities." *Proceedings of the National Academy of Sciences of the United States of America 99*(22): 14250-14255.

Brenner, S. and R. W. Horne (1959). "A negative staining method for high resolution electron microscopy of viruses." *Biochimica et biophysica acta 34*: 103-110.

Brenner, S., G. Streisinger, et al. (1959). "Structural components of bacteriophage." *Journal of Molecular Biology 1*(3): 281-IN221.

Brovko, L. Y., H. Anany, et al. (2012). "Bacteriophages for detection and control of bacterial pathogens in food and food-processing environment." *Adv Food Nutr Res 67*: 241-288.

Brussow, H. and R. W. Hendrix (2002). "Phage genomics: small is beautiful." *Cell 108*(1): 13-16.

Bruttin, A. and H. Brussow (2005). "Human volunteers receiving Escherichia coli phage T4 orally: a safety test of phage therapy." *Antimicrob Agents Chemother 49*(7): 2874-2878.

Carlton, R. M. (1999). "Phage therapy: past history and future prospects." *Archivum immunologiae et therapiae experimentalis 47*(5): 267-274.

Casas, V. and S. Maloy (2011). "Role of bacteriophage-encoded exotoxins in the evolution of bacterial pathogens." *Future Microbiol 6*(12): 1461-1473.

Chaignon, P., I. Sadovskaya, et al. (2007). "Susceptibility of staphylococcal biofilms to enzymatic treatments depends on their chemical composition." *Appl Microbiol Biotechnol 75*(1): 125-132.

Chavez, C. L. and M. P. Calos (2011). "Therapeutic applications of the PhiC31 integrase system." *Curr Gene Ther 11*(5): 375-381.

Coffey, B., S. Mills, et al. (2010). "Phage and their lysins as biocontrol agents for food safety applications." *Annu Rev Food Sci Technol 1*: 449-468.

Comeau, A. M., G. F. Hatfull, et al. (2008). "Exploring the prokaryotic virosphere." *Research in microbiology 159*(5): 306-313.

Corbin, B. D., R. J. McLean, et al. (2001). "Bacteriophage T4 multiplication in a glucose-limited Escherichia coli biofilm." *Can J Microbiol 47*(7): 680-684.

Costa, A. R., M. Henriques, et al. (2009). "The role of polysaccharide intercellular adhesin (PIA) in Staphylococcus epidermidis adhesion to host tissues and subsequent antibiotic tolerance." *Eur J Clin Microbiol Infect Dis 28*(6): 623-629.

Costerton, J. W., K. J. Cheng, et al. (1987). "Bacterial biofilms in nature and disease." *Annu Rev Microbiol 41*: 435-464.

Cvirkaite-Krupovic, V., M. M. Poranen, et al. (2010). "Phospholipids act as secondary receptor during the entry of the enveloped, double-stranded RNA bacteriophage phi6." *The Journal of general virology 91*(Pt 8): 2116-2120.

D'Herelle, F. (2007). "On an invisible microbe antagonistic toward dysenteric bacilli: brief note by Mr. F. D'Herelle, presented by Mr. Roux. 1917." *Research in microbiology 158*(7): 553-554.

Debarbieux, L., D. Leduc, et al. (2010). "Bacteriophages can treat and prevent Pseudomonas aeruginosa lung infections." *J Infect Dis 201*(7): 1096-1104.

Donlan, R. M. (2009). "Preventing biofilms of clinically relevant organisms using bacteriophage." *Trends Microbiol 17*(2): 66-72.

Duckworth, D. (1987). "History and basic properties of bacterial viruses." In: *Phage Ecology* (Goyal, S.M., Gerba, C.P. and Bitton, G., Eds.), pp. 1–44. John Wiley & Sons, New York.

Eisenstark, A. (1967). Bacteriophage techniques. In: Maramorosch K, Koprowski H (eds), *Methods in Virology* Vol *1*. Academic Press, NewYork, pp 449–525.

Fiorentin, L., N. D. Vieira, et al. (2005). "Oral treatment with bacteriophages reduces the concentration of Salmonella Enteritidis PT4 in caecal contents of broilers." *Avian Pathol 34*(3): 258-263.

Flemming, H. C. and J. Wingender (2010). "The biofilm matrix." *Nat Rev Microbiol 8*(9): 623-633.

Friedman, D. I. and D. L. Court (2001). "Bacteriophage lambda: alive and well and still doing its thing." *Current opinion in microbiology 4*(2): 201-207.

Fu, W., T. Forster, et al. (2010). "Bacteriophage cocktail for the prevention of biofilm formation by Pseudomonas aeruginosa on catheters in an in vitro model system." *Antimicrob Agents Chemother 54*(1): 397-404.

Funatsu, T., T. Taniyama, et al. (2002). "Rapid and sensitive detection method of a bacterium by using a GFP reporter phage." *Microbiol Immunol 46*(6): 365-369.

Gast, R. K., R. E. Porter, Jr., et al. (1997). "Assessing the sensitivity of egg yolk antibody testing for detecting Salmonella enteritidis infections in laying hens." *Poult Sci 76*(6): 798-801.

Geier, M. R., M. E. Trigg, et al. (1973). "Fate of bacteriophage lambda in non-immune germ-free mice." *Nature 246*(5430): 221-223.

Goodridge, L., J. Chen, et al. (1999). "The use of a fluorescent bacteriophage assay for detection of Escherichia coli O157:H7 in inoculated ground beef and raw milk." *Int J Food Microbiol 47*(1-2): 43-50.

Gottesman, M. E. and R. A. Weisberg (2004). "Little lambda, who made thee?" *Microbiology and molecular biology reviews : MMBR 68*(4): 796-813.

Griffin, D. W., C. J. Gibson, 3rd, et al. (1999). "Detection of viral pathogens by reverse transcriptase PCR and of microbial indicators by standard methods in the canals of the Florida Keys." *Appl Environ Microbiol 65*(9): 4118-4125.

Hagens, S. and M. J. Loessner (2007). "Application of bacteriophages for detection and control of foodborne pathogens." *Appl Microbiol Biotechnol 76*(3): 513-519.

Hankin, E. H. (1896). L'action bactéricide des eaux de la Jumna et du Gange sur le microbe du choléra. *Ann. Inst. Pasteur 10*:511-523.

Hanlon, G. W., S. P. Denyer, et al. (2001). "Reduction in exopolysaccharide viscosity as an aid to bacteriophage penetration through Pseudomonas aeruginosa biofilms." *Appl Environ Microbiol 67*(6): 2746-2753.

Hatfull, G. F. and R. W. Hendrix (2011). "Bacteriophages and their genomes." *Current opinion in virology 1*(4): 298-303.

Hatfull, G. F., M. L. Pedulla, et al. (2006). "Exploring the mycobacteriophage metaproteome: phage genomics as an educational platform." *PLoS genetics 2*(6): e92.

Hendrix, R. W., M. C. Smith, et al. (1999). "Evolutionary relationships among diverse bacteriophages and prophages: all the world's a phage." *Proceedings of the National Academy of Sciences of the United States of America 96*(5): 2192-2197.

Jado, I., R. Lopez, et al. (2003). "Phage lytic enzymes as therapy for antibiotic-resistant Streptococcus pneumoniae infection in a murine sepsis model." *J Antimicrob Chemother 52*(6): 967-973.

Juhala, R. J., M. E. Ford, et al. (2000). "Genomic sequences of bacteriophages HK97 and HK022: pervasive genetic mosaicism in the lambdoid bacteriophages." *Journal of Molecular Biology 299*(1): 27-51.

King, A. M. Q., M. J. Adams, et al. (2012). Virus taxonomy: classification and nomenclature of viruses. *Ninth Report of the International Committee on Taxonomy of Viruses.* Elsevier–Academic Press, London, (in print, ISBN: 978-0-12- 384684-6).

Kobiler, O., A. Rokney, et al. (2005). "Quantitative kinetic analysis of the bacteriophage lambda genetic network." *Proceedings of the National Academy of Sciences of the United States of America 102*(12): 4470-4475.

Kropinski, A. M. (2006). "Phage Therapy - Everything Old is New Again." *The Canadian journal of infectious diseases & medical microbiology = Journal canadien des maladies infectieuses et de la microbiologie medicale / AMMI Canada 17*(5): 297-306.

Krupovic, M., D. Prangishvili, et al. (2011). "Genomics of bacterial and archaeal viruses: dynamics within the prokaryotic virosphere." *Microbiology and molecular biology reviews : MMBR 75*(4): 610-635.

Kutateladze, M. and R. Adamia (2010). "Bacteriophages as potential new therapeutics to replace or supplement antibiotics." *Trends Biotechnol 28*(12): 591-595.

Kwan, T., J. Liu, et al. (2005). "The complete genomes and proteomes of 27 Staphylococcus aureus bacteriophages." *Proceedings of the National Academy of Sciences of the United States of America 102*(14): 5174-5179.

Kwan, T., J. Liu, et al. (2006). "Comparative genomic analysis of 18 Pseudomonas aeruginosa bacteriophages." *Journal of bacteriology 188*(3): 1184-1187.

Lazareva, E. B., S. V. Smirnov, et al. (2001). "[Efficacy of bacteriophages in complex treatment of patients with burn wounds]." *Antibiot Khimioter 46*(1): 10-14.

Lederberg, J. (1996). "Smaller fleas ... ad infinitum: therapeutic bacteriophage redux." *Proc Natl Acad Sci U S A 93*(8): 3167-3168.

Leverentz, B., W. S. Conway, et al. (2001). "Examination of bacteriophage as a biocontrol method for salmonella on fresh-cut fruit: a model study." *J Food Prot 64*(8): 1116-1121.

Leverentz, B., W. S. Conway, et al. (2003). "Biocontrol of Listeria monocytogenes on fresh-cut produce by treatment with lytic bacteriophages and a bacteriocin." *Appl Environ Microbiol 69*(8): 4519-4526.

Lima-Mendez, G., A. Toussaint, et al. (2011). "A modular view of the bacteriophage genomic space: identification of host and lifestyle marker modules." *Research in microbiology 162*(8): 737-746.

Loeffler, J. M., S. Djurkovic, et al. (2003). "Phage lytic enzyme Cpl-1 as a novel antimicrobial for pneumococcal bacteremia." *Infect Immun 71*(11): 6199-6204.

Loeffler, J. M., D. Nelson, et al. (2001). "Rapid killing of Streptococcus pneumoniae with a bacteriophage cell wall hydrolase." *Science 294*(5549): 2170-2172.

Loessner, M. J., A. Schneider, et al. (1996). "Modified Listeria bacteriophage lysin genes (ply) allow efficient overexpression and one-step purification of biochemically active fusion proteins." *Appl Environ Microbiol 62*(8): 3057-3060.

Lopez, R., E. Garcia, et al. (1997). "The pneumococcal cell wall degrading enzymes: a modular design to create new lysins?" *Microb Drug Resist 3*(2): 199-211.

Lukacik, P., T. J. Barnard, et al. (2012). "Using a bacteriocin structure to engineer a phage lysin that targets Yersinia pestis." *Biochem Soc Trans 40*(6): 1503-1506.

Lukacik, P., T. J. Barnard, et al. (2012). "Structural engineering of a phage lysin that targets gram-negative pathogens." *Proc Natl Acad Sci U S A 109*(25): 9857-9862.

Ma, Y., J. C. Pacan, et al. (2008). "Microencapsulation of bacteriophage felix O1 into chitosan-alginate microspheres for oral delivery." *Appl Environ Microbiol 74*(15): 4799-4805.

Markoishvili, K., G. Tsitlanadze, et al. (2002). "A novel sustained-release matrix based on biodegradable poly(ester amide)s and impregnated with bacteriophages and an antibiotic shows promise in management of infected venous stasis ulcers and other poorly healing wounds." *Int J Dermatol 41*(7): 453-458.

Matinkhoo, S., K. H. Lynch, et al. (2011). "Spray-dried respirable powders containing bacteriophages for the treatment of pulmonary infections." *J Pharm Sci 100*(12): 5197-5205.

Merabishvili, M., J. P. Pirnay, et al. (2009). "Quality-controlled small-scale production of a well-defined bacteriophage cocktail for use in human clinical trials." *PLoS One 4*(3): e4944.

Merril, C. R., B. Biswas, et al. (1996). "Long-circulating bacteriophage as antibacterial agents." *Proceedings of the National Academy of Sciences of the United States of America 93*(8): 3188-3192.

Merril, C. R., D. Scholl, et al. (2003). "The prospect for bacteriophage therapy in Western medicine." *Nat Rev Drug Discov 2*(6): 489-497.

Nelson, D., L. Loomis, et al. (2001). "Prevention and elimination of upper respiratory colonization of mice by group A streptococci by using a bacteriophage lytic enzyme." *Proc Natl Acad Sci U S A 98*(7): 4107-4112.

Nishikawa, H., M. Yasuda, et al. (2008). "T-even-related bacteriophages as candidates for treatment of Escherichia coli urinary tract infections." *Arch Virol 153*(3): 507-515.

Perepanova, T. S., O. S. Darbeeva, et al. (1995). "[The efficacy of bacteriophage preparations in treating inflammatory urologic diseases]." *Urol Nefrol (Mosk)*(5): 14-17.

Pope, W. H., D. Jacobs-Sera, et al. (2011). "Expanding the diversity of mycobacteriophages: insights into genome architecture and evolution." *PloS one 6*(1): e16329.

Ptashne, M. A. (1992). Genetic Switch: Phage Lambda and Higher Organisms. Blackwell Publ.

Revet, B., B. von Wilcken-Bergmann, et al. (1999). "Four dimers of lambda repressor bound to two suitably spaced pairs of lambda operators form octamers and DNA loops over large distances." *Current biology : CB 9*(3): 151-154.

Ricard, J. D. (2012). "New therapies for pneumonia." *Curr Opin Pulm Med 18*(3): 181-186.

Robert, M. A., Y. Zeng, et al. (2012). "Efficacy and site-specificity of adenoviral vector integration mediated by the phage PhiC31 integrase." *Hum Gene Ther Methods*.

Schofield, D. A., N. J. Sharp, et al. (2012). "Phage-based platforms for the clinical detection of human bacterial pathogens." *Bacteriophage 2*(2): 105-283.

Schuch, R., D. Nelson, et al. (2002). "A bacteriolytic agent that detects and kills Bacillus anthracis." *Nature 418*(6900): 884-889.

Shapiro, O. H. and A. Kushmaro (2011). "Bacteriophage ecology in environmental biotechnology processes." *Current opinion in biotechnology 22*(3): 449-455.

Sharma, M., J. R. Patel, et al. (2009). "Effectiveness of bacteriophages in reducing Escherichia coli O157:H7 on fresh-cut cantaloupes and lettucet." *J Food Prot 72*(7): 1481-1485.

Sillankorva, S., D. Oliveira, et al. (2011). "Efficacy of a broad host range lytic bacteriophage against E. coli adhered to urothelium." *Curr Microbiol 62*(4): 1128-1132.

Sillankorva, S., E. Pleteneva, et al. (2010). "Salmonella Enteritidis bacteriophage candidates for phage therapy of poultry." *J Appl Microbiol 108*(4): 1175-1186.

Snyder, L., W. Champness (2007). Molecular genetics of bacteria. 3rd edition. ASM Press. Washington, D. C.

Sulakvelidze, A., Z. Alavidze, et al. (2001). "Bacteriophage therapy." *Antimicrobial agents and chemotherapy 45*(3): 649-659.

Sulakvelidze A. Z. et al. (2009). "*Pseudomonas aeruginosa*: bacteriophage and uses thereof". United States Patent: 7,622,293, 2009

Suttle, C. A. (2005). "Viruses in the sea." *Nature 437*(7057): 356-361.

Svenningsen, S. L., N. Costantino, et al. (2005). "On the role of Cro in lambda prophage induction." *Proceedings of the National Academy of Sciences of the United States of America 102*(12): 4465-4469.

Tsao et al. (2012) "Newly isolated bacteriophage specific to *Klebsiella pneumoniae*." United States Patent8,252,575.

Toro, H., S. B. Price, et al. (2005). "Use of bacteriophages in combination with competitive exclusion to reduce Salmonella from infected chickens." *Avian Dis 49*(1): 118-124.

Twort, F. W. (1915). "AN INVESTIGATION ON THE NATURE OF ULTRA-MICROSCOPIC VIRUSES." *The Lancet 186*(4814): 1241-1243.

Ulitzur, S. and J. Kuhn (2000). "Construction of lux bacteriophages and the determination of specific bacteria and their antibiotic sensitivities." *Methods Enzymol 305*: 543-557.

Vandersteegen, K., W. Mattheus, et al. (2011). "Microbiological and molecular assessment of bacteriophage ISP for the control of Staphylococcus aureus." *PLoS One 6*(9): e24418.

Venema, G., J. Kok, et al. (1999). "From DNA sequence to application: possibilities and complications." *Antonie van Leeuwenhoek 76*(1-4): 3-23.

Vieira, A., Y. J. Silva, et al. (2012). "Phage therapy to control multidrug-resistant Pseudomonas aeruginosa skin infections: in vitro and ex vivo experiments." *Eur J Clin Microbiol Infect Dis 31*(11): 3241-3249.

Waldor, M. K. F., D. I.; Adhya, S. L. (2005). *Phages: Their roles in bacterial pathogenesis and biotechnology*. Washington, DC, ASM Press.

Weinbauer, M. G. (2004). "Ecology of prokaryotic viruses." *FEMS microbiology reviews 28*(2): 127-181.

Westwater, C., L. M. Kasman, et al. (2003). "Use of genetically engineered phage to deliver antimicrobial agents to bacteria: an alternative therapy for treatment of bacterial infections." *Antimicrobial agents and chemotherapy 47*(4): 1301-1307.

Wright, A., C. H. Hawkins, et al. (2009). "A controlled clinical trial of a therapeutic bacteriophage preparation in chronic otitis due to antibiotic-resistant Pseudomonas aeruginosa; a preliminary report of efficacy." *Clin Otolaryngol 34*(4): 349-357.

In: Bacteriophages
Editors: Clark Denton and Richard J. Crosby
ISBN: 978-1-62618-513-5

Chapter 3

Abortive Infection: A Killer Bacteriophage Resistance Strategy of Last Resort

***Chidiebere Akusobi*[†1], *Feng Rao*[†1], *Francesca L. Short*[1] *and George P. C. Salmond*[1*]**
[1]Department of Biochemistry, University of Cambridge, Cambridge, UK

Abstract

Bacteriophages are the most abundant biological entities on earth and outnumber their bacterial hosts by a factor of 10. As a result, bacteria have evolved numerous defense systems to resist viral infection. These systems operate at all stages of the phage life cycle and include cell surface modification, restriction-modification and CRISPR-mediated resistance. Another notable defense strategy is phage abortive infection (Abi), where a bacterial host prevents the production of viable phage progeny and in the process protects the bacterial population as a whole, but at the expense of its life. For this reason, phage Abi can be viewed as "altruistic suicide". Recently, a link has been demonstrated between abortive infection activity and the near-ubiquitous prokaryotic toxin-antitoxin (TA) systems. Here, we review abortive infection and the emerging role for TA systems in mediating this powerful anti-viral activity.

Introduction

Bacteriophages are obligate parasites that require bacterial hosts to replicate. Phages initiate infections by first attaching to the surface of bacteria and injecting their genomes into host cells. Virulent phages then hijack the host's cellular machinery to direct the production of phage virions that burst from the cell leading to bacterial lysis. In contrast, temperate

† These authors contributed equally to the work.
* To whom correspondence should be addressed: gpcs2@cam.ac.uk.

phages insert their genomes into the bacterial chromosome or replicate as plasmids with the host. Environmental stress can induce the quiescent genome, known as a prophage, to enter the lytic cycle and produce infectious viral particles. In both life cycles, successful infection leads to death of the host and production of more phage that go on to infect susceptible cells (Figure 1).

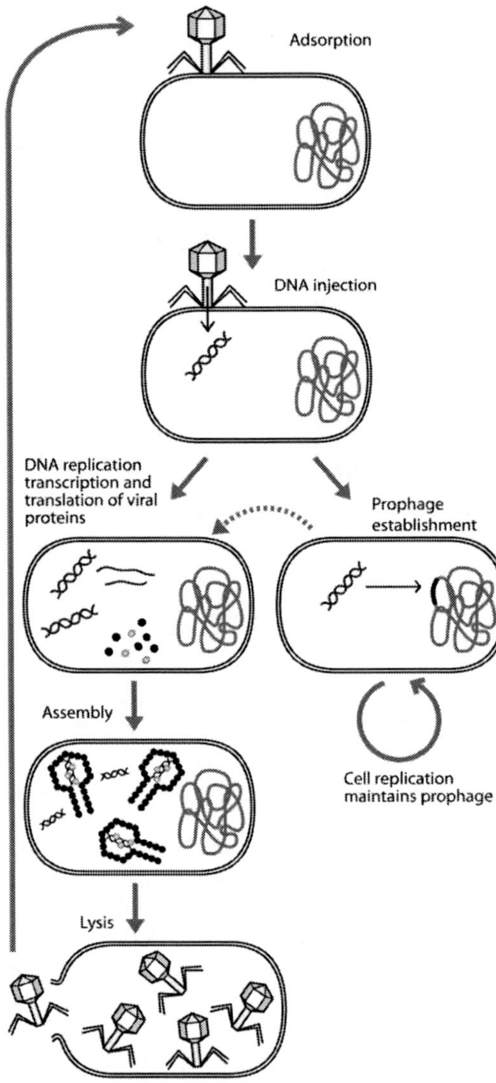

Figure 1. The phage life cycle. Phage particles must first attach to the surface of the bacterial cell (adsorption). The genome is then injected and in the lytic pathway (left), drives the transcription and translation of phage products as well as its own replication. The phage products assemble into new virions and finally lyse the host cell to release the progeny. Alternatively (right), following DNA injection the phage genome can integrate into the bacterial chromosome to form a prophage. The prophage is replicated along with the host cell until an external trigger causes it to re-enter the lytic pathway. Note that the genomes of some temperate phages replicate as plasmids within their respective hosts.

Despite preying on bacteria, phages do not always have a strictly antagonistic relationship with their hosts. They contribute to bacterial evolution by facilitating horizontal gene transfer. During infection, some phages can accidentally package host DNA and transfer this "foreign" DNA to a new cell, where it can be incorporated into the chromosome of the recipient. This process, known as transduction, increases the genetic diversity of bacteria and can be useful when adapting to new or changing environments [1]. Phages also play key roles in bacterial pathogenesis. Many virulence factors are encoded by prophages; notable examples include shiga toxins in enterohemorrhagic *Escherichia coli* and cholera toxin in *Vibrio cholerae* [2]. In lysogenic conversion, prophage infection can convert non-pathogenic bacteria into a pathogenic strain. A classic example of this process occurs in *Corynebacterium diphtheriae*, which produces a diphtheria toxin encoded by the β^{tox+} prophage. Bacteria with the incorporated prophage are known as lysogens and these "converted" hosts are responsible for diphtheria infections [3, 4]. Prophage-encoded virulence factors can allow bacteria to penetrate, colonize, and survive in new organisms. As a result, prophage-encoded genes have become important in the genomes of some pathogenic bacteria due to the adaptive advantages they confer. Thus, through both transduction and lysogenic conversion, phages provide new functions to bacteria and thereby can contribute significantly to the evolution and diversity of their hosts [1].

As obligate parasites, phages are important determinants of bacterial adaptive evolution. Phages are the most abundant biological entities on Earth; with an estimated 10^{31} phage particles, they outnumber bacteria by a factor of 10 [5-7]. Furthermore, with an estimated 10^{25} bacterial virus infections occurring per second globally, there is no doubt that phage infection is a strong natural selective force for bacteria [6, 8]. Unsurprisingly, bacteria have evolved numerous resistance mechanisms to block phage attack, and in response, phage employ counter strategies that overcome bacterial defense mechanisms to ensure productive infections [9, 10]. This co-evolutionary "arms race" has been ongoing for millennia and is a major factor in the evolution of both bacteria and their viral parasites. Extensive research has been done to characterize this predator-prey dynamic, especially because the phage-host relationship is pertinent to several applications of phage biology. The first is phage therapy, a treatment that uses the lethal activity of phage to treat bacterial infections [11]. Researchers studying phage therapy are rightfully concerned with the emergence of phage resistance in the bacterial pathogens they are attempting to kill. Although multiple phages that target different host receptors are generally used in therapeutic cocktails (to diminish the impact of spontaneous phage resistance), understanding phage resistance mechanisms may allow the design of better phage therapeutics that delay or thwart the emergence of resistance [12]. The second application involving phage-host interactions is in industrial fermentation. The cheese-manufacturing industry uses lactic acid bacteria for fermentations, which are prone to phage infection - leading to economic losses. Significant numbers of milk fermentation cultures are affected every year prompting considerable research into *Lactococcus* phage resistance mechanisms has been conducted in order to better understand how to protect these cultures from viral infection [13-15].

Mechanisms of Bacteriophage Resistance

Blocking Phage Adsorption

The first step of phage infection is adsorption where the virus attaches to a surface receptor on the host. Hence, a common strategy bacteria use to resist phage attack is blocking phage adsorption. Bacteria do this in various ways; the first is by deleting or modifying the phage receptor. Phage receptors are often membrane proteins or proteinaceous structures (e.g. porins, pili, and flagella etc.) that usually perform important functions for the cell [16]. Consequently, bacteria must retain some phage receptors, but can prevent phage adsorption by modifying residues on the protein that alter its structure while retaining its biological function [17]. Such modifications can decrease the phage attachment rate or block adsorption completely. An example of this strategy is the mutation of some outer membrane proteins (Omp) in *Escherichia coli* that confer phage resistance. Some viruses use porins such as OmpA, OmpF, or OmpC as receptors, and certain mutations in the surface-exposed region of these proteins have been shown to confer resistance to phage adsorption whilst retaining porin function [16, 18, 19]. In addition, some phages of Gram-negative bacteria bind the LPS layer of the cell surface; mutation of these moieties has been shown to thwart phage adsorption [20-22]. Furthermore, downregulation of receptor expression or complete receptor loss can also block phage adsorption [23]. This was recently shown in a co-evolution experiment with phage λ and *E. coli* B. Meyer et al. found that the bacterial populations repeatedly evolved to downregulate receptor expression [24]. In other phage-host co-evolution experiments, researchers found that the initial phage receptor was no longer expressed due to deletion mutations [23, 25]. Altering the structure or expression of receptors may cause fitness costs for the bacteria via growth defects, and as a result, such mutations are often only advantageous in the presence of the phage [26].

Lastly, bacteria can block phage adsorption by concealing membrane receptors with extracellular matrices or membrane associated inhibitor proteins [27, 28]. An example of inhibitor proteins was discovered early on in *Staphylococcus aureus.* The bacterium encodes protein A, which binds to receptors and masks the sites where Φ52 and Φ80 would otherwise bind [29]. Moreover, extracellular matrices composed of lipopolysaccharides have been described in *Azotobacter* spp., *Pseudomonas* spp., and *E. coli* strains [30]. Matrices can serve dual purposes of preventing phage from binding to receptors and increasing bacterial survival in harsh environments [31].

To overcome the physical barrier of the extracellular matrix, some phages have evolved the ability to degrade extracellular capsules and allow access to membrane receptors using enzymes as structural components of the viral capsids [30, 32]. Furthermore, phage can evolve the ability to bind a mutated receptor or a new receptor if the original receptor is downregulated or lost [19, 33]. In the phage λ and *E. coli* B co-evolution experiment, four mutations in the virus' receptor recognition protein allowed the phage to bind OmpF instead of its native receptor, LamB, whose expression was downregulated [24].

Genome Uptake Blocks

Bacteria can resist phage infection by blocking the entry of the phage genome into the cell. With this strategy adsorption occurs normally, but injection is blocked [34, 35]. Genome uptake blocks have been best described in *Lactococcus lactis.* This bacterium encodes PIP (phage infection protein), which is required by several phages to transport their genomes into the cell. Phage resistant bacteria can harbor mutations in PIP that render them insensitive to phage infection by impeding DNA uptake [36]. In addition, the *Lactococcus* plasmid, pNP40, was found to confer resistance to Φc2 by blocking DNA uptake via an unknown mechanism [37]. Superinfection exclusion (Sie) is another mechanism that blocks genome injection, but is commonly mediated by a prophage. Sie systems encode proteins that block the uptake of phage genomes and thus protect the host from phage infection [38, 39]. In this manner, Sie can be described as a phage defense mechanism against competing phage infections, with the fortuitous consequence of granting resistance to the host. Sie is best characterized in T-even phages and their *E. coli* hosts. Coliphage T4 has two genes, *imm* and *sp*, that prevent the injection of T-even like phage DNA into hosts already infected with T4 [39, 40]. While Sie grants superinfection immunity to bacteria, prophages with Sie mechanisms can still be induced to enter thc lytic cycle and kill its host. When this occurs, bacteria in the population lacking a Sie encoding prophage are vulnerable to infection, whereas cells with the prophage are left unscathed [41].

Restriction-Modification

The phage genome, if successfully injected into the host, must evade bacterial systems that recognize and cleave foreign DNA. One of these mechanisms is the restriction-modification (R-M) system, which is ubiquitous among bacteria and effective against a wide range of phage. R-M systems are comprised of at least two proteins, a methyltransferase that methylates host DNA, designating it as "self," and a restriction endonuclease that cleaves unmodified DNA [42]. One class of R-M system has an additional protein that determines specificity for the endonuclease and methyltransferase (38). When unmodified phage DNA enters the cytoplasm, restriction enzymes recognize it as foreign and digest it at specific sites along the viral genome, thus blocking phage productive infection [42]. Four types of restriction modification systems have been discovered thus far. The Type I, II, and III systems use the conventional method of digesting unmodified DNA, while the Type IV system digests modified phage DNA [43]. An R-M-like system in *Streptomyces coelicolor*, known as phage growth limitation (Pgl), modifies phage DNA upon entry into the host. When this modified phage infects a cell with a Pgl system, it is targeted for destruction [44].

Phages evade R-M systems using several strategies that include losing or changing sequences recognized by endonucleases, incorporating modified nucleotides like hydroxymethylated bases into their genomes, and inactivating the restriction enzymes. Interestingly, some phage can encode their own methyltransferases to modify their DNA and thereby evade digestion [45-47].

CRISPR/Cas System

CRISPRs (clustered regularly interspaced short palindromic repeats) and Cas (CRISPR-associated) proteins are the only adaptive resistance mechanism in bacteria known to date. The other resistance strategies indiscriminately affect phage while the CRISPR/Cas system recognizes specific DNA sequences in phage genomes [48]. CRISPR/Cas systems were discovered recently and have been found in both bacteria and archaea [49]. The short palindromic repeats of DNA (24-47 base pairs) in CRISPRs flank segments of foreign DNA known as 'spacer regions' that are comprised of DNA from previously infecting phage and plasmids [48]. Little is known about how bacteria acquire the foreign DNA; but it is known that through the activity of Cas proteins, foreign phage DNA is recognized and targeted for degradation, thus granting the bacteria resistance [48]. For a more mechanistic review of the three types of CRISPR/Cas systems, see reference [50].

The CRISPR/Cas system is usually described as adaptive in bacteria since new spacer regions can be added to the system after an infection. As a result, bacteria gain immunity to subsequent infections from the same phage [51]. While the CRISPR/Cas system may be adaptive, it is not a foolproof method of resistance. Phage specific to *Streptococcus thermophilus* have been shown to evade the bacterium's CRISPR/Cas system by acquiring a mutation in a proto-spacer region, which is broadly defined as the region in the phage genome complementary to the spacer region in the CRISPR/Cas [50]. Phage specific to *Pseudomonas aeruginosa* have also been discovered with anti-CRISPR genes [52]. Given that CRISPR/Cas is widespread among bacteria, there are undoubtedly other strategies phage employ to combat this system that have not yet been discovered.

Abortive Infection

Abortive infection (Abi) impedes crucial stages of the phage life cycle like replication, transcription, and virion maturation, and in the process of aborting the phage infection, the infected cell dies. As a term, abortive infection refers to any mechanism that disrupts the phage life cycle despite successful adsorption, genome uptake, and evasion of R-M and CRISPR/Cas systems. Abortive infection is unique because it is the only resistance mechanism that leads to the death of both the phage and the host. In the literature, Abi is described as an example of altruistic suicide, for in killing itself, the bacterium is hypothesized to protect the rest of the population from phage infection. However, to date, no published work has confirmed experimentally that Abi is truly an example of microbial altruistic suicide.

Abortive infection systems have been described in a number of bacterial species and have been categorized extensively in lactic acid bacteria, where they are exploited by the fermentation industry to keep large volume bacterial cultures resistant to phage infection [53]. Although Abi systems work using different mechanisms, a commonality in all systems is the presence of a toxic protein that is activated upon phage infection. The identity of the toxic protein varies depending on the system; examples include endoribonucleases, membrane ion channels, and transcriptional repressors. The proteins, once activated, have bacteriostatic or bacteriolytic effects on the cell, which prevents propagation of the phage. Predictably, phage have evolved strategies to resist Abi systems. Most recently, ΦTE was found to evade the

ToxIN Abi mechanism of *Pectobacterium atrosepticum* by expressing an RNA antitoxin mimic that suppressed the toxic protein, ToxN [54].

Summary of Phage Resistance Mechanisms

The five resistance strategies, adsorption resistance, genome uptake blocks, R-M and CRISPR systems, and abortive infection are products of an ongoing co-evolutionary arms race between bacteria and phage. While the mechanisms all have the intended effect of stopping phage infection, they do so at different stages of the phage life cycle. This ultimately leads to varying consequences for both phage and the host. The first resistance mechanism, blocking phage adsorption, is in some ways the first line of defense for bacteria. If a bacterium lacks the necessary receptor for a phage, an infection cannot be initiated. If a phage is able to infect the bacterium, blocking adsorption keeps both the bacterium and phage alive, and the latter can go on to infect another cell. From the bacterial standpoint, blocking adsorption by altering the structure or expression of a receptor is a stochastic process arising from mistakes in DNA replication. Receptor mutants are selected for their ability to resist phage infection. Unlike adsorption blocks, resistance via preventing genome uptake kills the phage and leaves the bacterium unscathed.

In contrast to the stochastic nature of adsorption resistance, the R-M and CRISPR/Cas systems actively recognize and destroy phage DNA before it exerts control on cellular functions. These systems require a mechanism whereby the host recognizes DNA as foreign in order to kill the phage intracellularly. This can be viewed as a more active strategy than adsorption resistance or genome uptake blocks. Finally, among resistance mechanisms, abortive infection is in a class on its own. While it is still unclear how phage trigger the Abi mechanisms, because of their suicidal nature, the employment of an Abi strategy can be viewed as a high risk mechanism that must be tightly regulated as a last line of defense against phage infection. Even when a phage can bypass a host's resistance mechanisms, the virus may have reduced productivity through slowed adsorption, delayed lysis, or decreased burst size. Nevertheless, when phages succeed in establishing productive infections, they will inevitably encounter a mutant host that resists infection, thus fuelling the continual arms race between bacteria and bacteriophage.

Abortive Infection

An Altruistic Defense Strategy?

Abortive infection was first reported in 1954 by Samson Gross who noted that when starved *E. coli* K-12 cells were infected with bacteriophage T2, there was a discrepancy in the predicted mathematics of the first round of infection [55]. The difference between the total number of plaques (which form as a result of either individual free phage or the bursting infected bacteria) and the titer of free phage was substantially less than the number of infected bacteria. In short, some bacterial cells were being infected but not releasing viable progeny.

Since these initial observations, abortive infection has been reported in many other species of bacteria, including *Lactococcus lactis* [56], *Bacillus subtilis* [57], and *Shigella dysenteriae* [58]. The term has been used also in mammalian virology to describe the equivalent phenomenon. Of note is the diversity of abortive infection systems found in *L. lactis*, the bacteria used in the cheese and yogurt fermentation industry. Most of the cognate Abi system genes are retained in the lactococcal genome through artificial selection in order to preserve economic viability [13].

While this Abi phenomenon is widespread through bacterial genera, the best characterized abortive infection systems (Rex, Lit, PrrC) are those common to *E. coli*. Studies into these systems are reviewed comprehensively elsewhere [59], so the discussion in the present review will be brief and only to provide context for the more recent developments in the field.

The Rex, Lit, and PrrC Systems

Rex, Lit, and PrrC are systems with different cellular targets within the cell. Rex is believed to act on the membrane to discharge the proton motive force, Lit is a protease that cleaves EF-Tu, while PrrC is an endoribonuclease for a lysine tRNA (Figure 2). Within all of these systems, "hallmarks" of an abortive infection mechanism can be seen: each recognizes phage attack before activating the killing mechanism and each of these systems keeps the active component "inactive" or suppressed during normal bacterial growth.

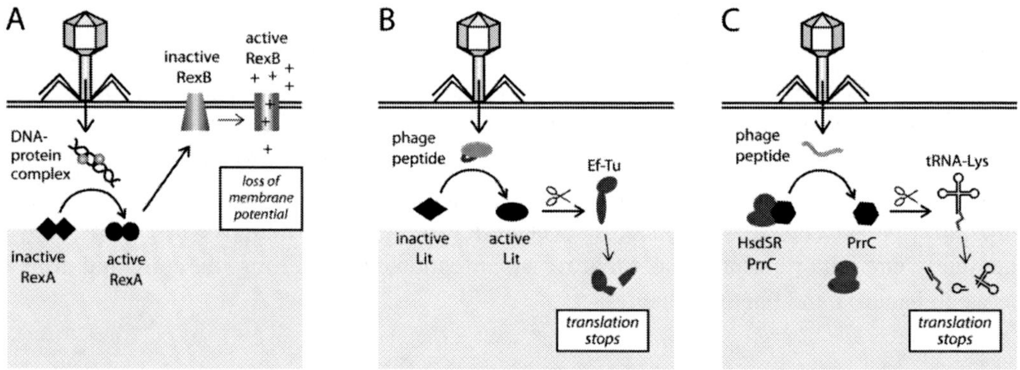

Figure 2. Three abortive infection systems of *E. coli*. *A* RecA is inert in the cell until activated by a complex of phage DNA with a bacterial protein. RecA then activates the membrane ion channel RecB, resulting in cell death due to depolarization of the cell membrane. *B* The Lit protein is activated by a specific phage peptide on a larger protein. Activated Lit cleaves the translation elongation factor Ef-Tu, resulting in cell death due to translation arrest. *C* The PrrC abortive infection protein is normally held in an inactive complex by HsdSR, but this interaction is disrupted by a phage peptide. Free PrrC cleaves charged tRNA-Lys, which causes cell stasis and eventually death due to translation arrest.

Rex is a two-component system of RexA and RexB. RexA acts as the sensor in this system and is activated by the inevitable production of a phage protein-bacterial DNA complex during infection of *E. coli* by phage T4 [60]. Activated RexA will in turn activate RexB, hypothesized, through modeling, to be a membrane ion channel. Activated RexB dissipates the proton motive force, which is required for both cellular ATP production as well as viral replication. Phage T4 was only highly susceptible to the Rex system with mutations in the rIIA and rIIB genes. A mutation in *motA*, which encodes a transcription factor, also allows T4 to bypass Rex completely [61].

Lit (Late Inhibitor of T4) is a protein encoded by a defective prophage e14, present in most K-12 derivatives of *E. coli* [62]. It is a zinc-dependent metalloprotease that cleaves a specific site on EF-Tu [63]. Lit is activated late in phage infection during which a peptide – the first 29 amino acids of the major head protein of the phage (also known as the Gol peptide) – becomes exposed to the cytoplasm. Ef-Tu is an essential component in the translation machinery, and so its cleavage and destruction will inevitably terminate phage replication, as well as the life of the host.

The *prrC* gene, sandwiched within a Type I R-M system (*hsdSR*), was found in a clinical isolate of *E. coli* – CT196. As mentioned above, the gene product PrrC acts as an endoribonuclease and like Lit, is activated by a small phage peptide from T4, Stp, which is not part of a larger protein product [64]. Rather than direct activation, PrrC is thought to be kept inactive by the HsdSR R-M complex. Stp interacts with the complex and renders it unable to sequester PrrC, leading to the cleavage of charged lysine tRNAs and the subsequent inhibition of the translation machinery [65]. The PrrC mediated abortive infection phenotype is only seen in infections by T4 phage mutants with mutations in either the polynucleotide kinase or the DNA ligase genes because these enzymes are able to repair the damage done by PrrC [66].

The bacteriophages that the PrrC and Rex systems affect contain a loss of function mutation that is critical for the abortive infection phenotype. This makes the ecological relevance of the systems questionable. The PrrC system, however, remains the only example of an abortive infection system that is intimately associated with a different, independently active, phage resistance locus – the HsdSR R-M system. HsdSR acts as the first barrier for phages; if they bypass this by disrupting the R-M complex, the host implements the abortive infection strategy. This suggests that the true role for this toxic ribonuclease is to confer phage resistance, and as result, the PrrC mechanism is often considered as evidence for the altruistic nature of abortive infection [59].

Other phage resistance mechanisms described here, such as R-M and CRISPR/Cas are always present and active within the cell. Abortive infection systems, however, cannot afford to be so indiscriminate because the active component is lethal. Hence, the host must have a mechanism to detect invading phage before activating the toxin. This additional complication makes Abi systems inherently difficult to study because researchers have to not only deal with the nature and function of the active component, but also with its regulation and suppression during the normal growth of the host.

Abi Systems in Lactococcus Lactis

As mentioned above, phage resistance mechanisms have been of great interest in bacteria used for industrial fermentation processes. In *L. lactis*, over 23 genetically distinct Abi systems have been found [13], and arguably these systems largely define the general foundation of modern research into abortive infection. However, because of a lack of homology between the proteins involved in these diverse systems with other proteins of known function, the molecular mechanisms of these systems are largely unknown. Therefore, the classic Lit, PrrC, and Rex systems remain the most "characterized".

The continuing research into the numerous Abi systems in *L. lactis* has focused on how they impede the infection process. When combined with the discovery of phage genes involved in sensitivity to Abi systems (presumably through direct or indirect interaction with the toxic protein), this information helps to elucidate mechanistic detail. For example, AbiD1 encodes a toxic protein, which is repressed in *L. lactis* under normal growth conditions. However, certain lactococcal phages produce a protein that induces the expression of AbiD1 during infection. This protein then binds RuvC-like nucleases from the phage and prevents the resolution of branched DNA structures essential for the replication and packaging of the phage genome [67]. Another system, AbiV, also produces a protein that binds to a phage peptide (SaV), and by an unknown mechanism, this complex prevents the translation of phage proteins in the later stages of infection [68].

A Link between Abortive Infection Systems and Toxin-antitoxin Systems

Toxin-antitoxin (TA) systems are genetic loci found ubiquitously in bacterial species and some species of archaea. They are known to play a role in the formation of persister cells and in responsiveness to environmental stresses [69]. Despite the division of TA systems into separate classes (Type I – Type V), central to all is the existence of a proteinaceous toxin which, if synthesized and active, becomes detrimental to the viability of the host cell (Table 1). Based on their physiological activities, it seems plausible that these systems are potential effectors of some abortive infection mechanisms.

A Type I TA locus (*hok/sok*) and a Type II TA locus (MazEF) have been reported to provide a detectable level of phage resistance [70, 71]. The Type II locus RnlAB has also been shown, in *E. coli*, to restrict T4 phage with a mutation in the *dmd* gene. This gene is a phage encoded antitoxin to the RnlA toxin which would otherwise kill the cell during phage infection [72]. Most recently, a Type III locus, ToxIN from *P. atrosepticum* has been shown to act as an abortive infection system, and arguably has become the best-studied abortive infection system mediated by a TA locus [73, 74]. As well as being extremely potent in excluding phage, an important feature compared to the other systems mentioned is that ToxIN is able to exclude a variety of phage in many bacterial species [73].

The relationship between TA systems and abortive infection brings the story back full circle to the original paper on abortive infection by Gross. He suspected that the *E. coli* physiology was related to the abortive infection phenotype. Specifically, he noted that starved cells showed phage resistance while normal growing cells were susceptible to the T2 phage [55]. However, when starved cells were immediately transferred to nutrient rich media at the point of infection, they did not revert back to a susceptible state. Thus, nutritional stress was deemed responsible for "abortive infection".

Table 1. Five Families of Toxin-Antitoxin Systems

Type I

Description: RNA antitoxin is complementary to toxin transcript. Binding causes degradation of the toxin mRNA[1]

Roles: Phage exclusion [70], plasmid stabilisation[2]

Distribution: Plasmids and chromosomes, vertically transmitted[3]

Type II

Description: Protein antitoxin inhibits protein toxin by direct interaction

Roles: Persistence[4], plasmid stabilisation[5], phage exclusion [71, 72], stress responses[6]

Distribution: Near-ubiquitous in bacterial and archaeal genomes, often present in very high numbers. Appear to be horizontally transferred between hosts[7]

Type III

Description: Processed RNA antitoxin inhibits protein toxin by direct interaction [73, 75]

Roles: Phage exclusion [73], plasmid stabilization [77]

Distribution: Plasmids and chromosomes, prevalent in *Firmicutes* with up to five per genome, horizontally transferred between hosts [76]

Type IV

Description: Protein antitoxin counteracts activity of protein toxin by reversing the toxin effect on its target, without antitoxin-toxin interaction[8]

Roles and distribution: Proposed role in cell survival under stress conditions[8], distribution not studied

Type V

Description: Protein antitoxin is a ribonuclease that cleaves toxin transcript[9]

Roles and distribution: Promotes formation of persister cells[9], distribution not studied

[1] Fozo EM, Hemm MR, Storz G (2008) Small toxic proteins and the antisense RNAs that repress them. *Microbiol Mol Biol R 72:* 579–589.

[2] Gerdes K, Larsen JE, Molin S (1985) Stable inheritance of plasmid R1 requires two different loci. *J Bacteriol 161:* 292–298.

[3] Fozo EM, Makarova KS, Shabalina S, Yutin N, Koonin EV, Storz G (2010) Abundance of type I toxin-antitoxin systems in bacteria: searches for new candidates and discovery of novel families. *Nucleic Acids Res 38:* 3743–3759.

[4] Maisonneuve E (2011) Bacterial persistence by RNA endonucleases. *Proc Natl Acad Sci USA 108:* 13206–13211.

[5] Ogura T, Hiraga S (1983) Mini-F plasmid genes that couple host cell division to plasmid proliferation. *Proc Natl Acad Sci USA 80:* 4784–4788.

[6] Leplae R, Geeraerts D, Hallez R, Guglielmini J, Drèze P, Van Melderen L (2011) Diversity of bacterial type II toxin-antitoxin systems: a comprehensive search and functional analysis of novel families. *Nucleic Acids Res 39:* 5513–5525.

[7] Masuda H, Tan Q, Awano N, Wu K-P, Inouye M (2012) YeeU enhances the bundling of cytoskeletal polymers of MreB and FtsZ, antagonizing the CbtA (YeeV) toxicity in Escherichia coli. *Mol Microbiol 84:* 979–989.

[8] Wang X, Lord DM, Cheng H-Y, Osbourne DO, Hong SH, Sanchez-Torres V, Quiroga C, Zheng K, Herrmann T, Peti W, Benedik MJ, Page R, Wood TK (2012) A new type V toxin-antitoxin system where mRNA for toxin GhoT is cleaved by antitoxin GhoS. *Nat Chem Biol 8:* 855–861.

P. atrosepticum ToxIN and Phage Counter-evolution

ToxIN, a Type III Toxin-Antitoxin System that Confers Phage Resistance

ToxIN was the first example of a Type III toxin-antitoxin locus, found on a cryptic plasmid in a strain of *P. atrosepticum* [73]. The toxic open reading frame encoded a 19.7 kDa endoribonuclease, ToxN. Immediately upstream in the locus were five and a half repeats of a 36 nucleotide sequence, which was processed by ToxN into a non-translated RNA pseudoknot, ToxI. In the cell, processed ToxI binds directly to ToxN in a heterohexameric stoichiometry and forms the inactive toxin-antitoxin complex. The structure of this protein-RNA complex has been solved and shows ToxIN forming a triangular structure with the protein monomers occupying the vertices and the RNA strands occupying the edges [75].

Uninhibited ToxN was shown to be extremely toxic to the cell due to cleavage of host mRNA. While this in itself was fascinating biology, the presence of this locus allowed the plasmid to confer phage resistance by abortive infection. In addition, when ToxN is heterologously expressed in other species, the same effect is seen against a spectrum of phages. ToxIN is a far more effective resistance locus than either *hok/sok* or MazEF, which were the first examples of toxin-antitoxin mediated phage exclusion. The *hok/sok* and MazEF systems were shown to reduce the efficiency of plaquing by up to 2-fold and around 500-fold respectively while ToxIN was effective up to 10^9-fold [70, 71, 73]. At these dramatic levels, in addition to a reduction in the number of infectious particles produced, the phage invader is effectively eradicated by the action of these suicide systems hiding within the bacteria they are unfortunate enough to infect.

This example highlights the difficulties in studying these mechanisms in depth. While the locus gives a clear phenotype to study and the mechanism of cell death is known, the question of how particular phage infection activates the system is not known. Indeed, the efficiency of the system at excluding phages means that it can be difficult to isolate so called "escape" mutant phages derived from originally susceptible phage. These "escapes" are useful for the investigation of viral genetic changes that render the phage immune to the system assuming that the relevant mutations in the "escape" phage are in genes involved with the activation of the dormant toxin-antitoxin system.

ΦTE Exemplifies a Way for a Phage to Protect Itself against ToxIN Mediated Abortive Infection

While ToxIN affected a variety of phages across bacterial genera, there are phages that completely ignored the system and did not induce abortive infection. Nevertheless, it is not unreasonable to assume that there will always be a system to which any particular phage is sensitive. Equally, in order for phages to survive, there must be a way to protect against this clearly effective bacterial defense strategy.

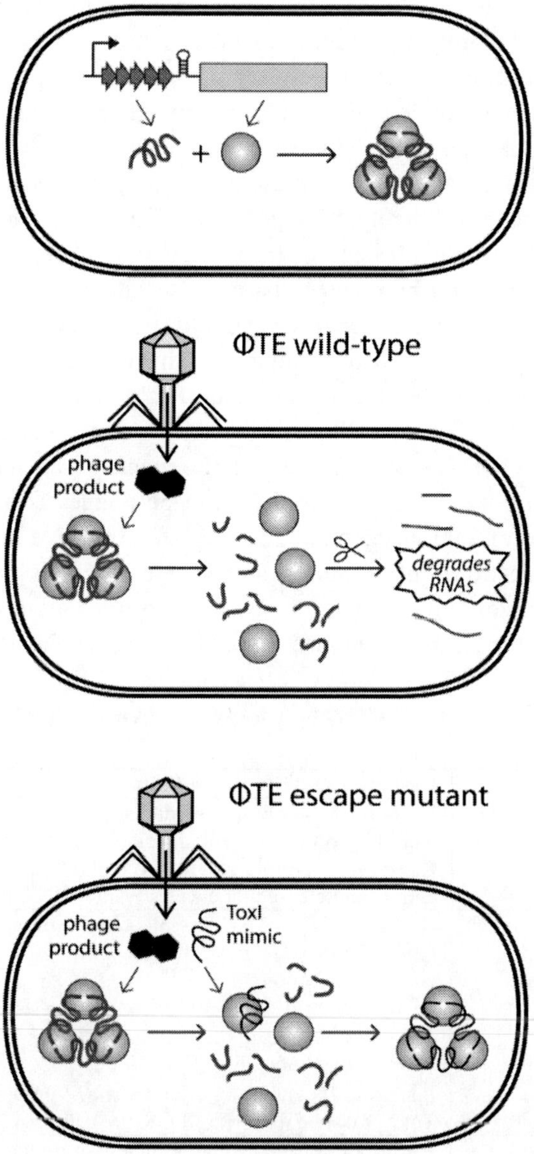

Figure 3. The abortive infection system *toxIN* comprises a series of 5.5 tandem repeats of the antitoxic RNA, ToxI, together with the gene for the ribonuclease ToxN. ToxN cleaves the repetitive ToxI transcript, and then is sequestered by ToxI in a triangular complex (top). Following infection with phage, such as ΦTE, an unknown phage product disrupts the complex, and free ToxN then cleaves cellular RNAs (centre). Escape mutants of ΦTE encode a mimic of ToxI, which prevents ToxN-mediated abortive infection by sequestering or suppressing the protein.

Very recently, one example of a mechanism by which phage are able to replicate in a cell "unnoticed" by ToxIN was uncovered, and it was attributed to the strategy of molecular mimicry [54]. ΦTE is a flagellum-dependent phage of *P. atrosepticum*, and the wild type phage has its efficiency of plaque formation attenuated by 10^8 fold by the ToxIN system present in the bacteria. However, occasionally, some plaques were seen on a lawn of bacteria containing the system and these phages were shown to evade the ToxIN system, even though they were derived from the progenitor susceptible ΦTE.

On analyzing the genetic differences between the ΦTE escapes and the wild type, Blower et al. noted that all the mutations were in a locus where the wild type phage contained a sequence similar, but not identical, to the sequence of the ToxI repeat motif in the ToxIN locus. In the case of the escape phages, they had undergone genetic amplification of this "pseudo-ToxI" sequence, presumably through replication errors. In an *in vivo* assay, the authors demonstrated that the expression of these pseudo-ToxI repeats in the phage acted to suppress ToxN during infection by phages that were sensitive to the system under "normal" conditions (Figure 3). One particular escape mutant, ΦTE-F, was especially interesting; rather than amplifying its own sequence, it had "hijacked" the ToxI repeats from the ToxIN locus through homologous recombination. This goes even further than just RNA mimicry – it is a remarkable example of the phage "stealing" the bacterium's own defense mechanism.

While the similarity between the pseudo-ToxI sequence and the natural antitoxin was serendipitous, the evolution of the various ΦTE escape phages is a clear demonstration of how the basic principles of natural selection allow phage to survive under the severe selective pressure of abortive infection. Furthermore, molecular mimicry is only one of a number of possible strategies phages might use to evade abortive infection. Some have been documented, while other plausible strategies can be hypothesized (Table 2).

Table 2. Mechanisms whereby phage can evade abortive infection

Mechanism of evasion	Rationale and literature example
Undo damage caused by toxic protein	**Rationale**: Mending the damage caused by the toxic Abi protein would keep the host cell alive and lead to a productive infection. **Example**: Coliphage T4 mends the damage caused by an endoribonuclease that cleaves lysine-tRNA in *E. coli* (66).
Encode protein that mimics activity of antitoxin or repressor of toxin protein	**Rationale**: Encoding a molecular mimic of a toxin's antitoxin or repressor would prevent the toxin from harming the host, thus ensuring infection. **Example**: ΦTE mutants encoded a pseudo-antitoxin of ToxN, an endoribonuclease toxin. Normally ToxN is inactivated by ToxI, but upon phage infection, ToxI is degraded and ToxN causes abortive infection. By encoding a ToxI mimic, ΦTE mutants prevent functional ToxN from being released into the cell (54).
Evade detection by Abi system	**Rationale**: Phage can either delete or mutate the gene that encodes the protein which triggers the abortive infection mechanism. In doing so, the virus would propagate in the host unrecognized. **Example**: Mutations in the *sav* gene of phage p2 confers resistance to AbiV of *Lactococcus*. The authors hypothesized that the gene product is involved in conferring sensitivity to AbiV (68).
Inactivate Abi toxin or suppress pathway	**Rationale**: Phage can encode a protein that inactivates the toxin and represses the Abi pathway. This method differs from molecular mimicry. In this case, the inactivating protein bears no similarity to the Abi antitoxin. **Example**: Coliphage T4 was shown to suppress the *lsoAB* TA system in *E. coli* O157:H7 by expressing Dmd. The protein acts like an antitoxin by binding and negating the activity of the LsoA toxin (72).
Enter lysogeny state	**Rationale**: A phage could enter lysogeny and evade detection by the host through avoiding independent replication.

The Future of Abortive Infection

Abortive infection is undoubtedly a powerful mechanism for the exclusion or control of phage parasites in bacterial populations. The diversity of mechanisms exemplified by the range of Abi systems in *Lactoccocus* is a sign that the genetic warfare between bacteria and phages has been ongoing with both sides evolving new mechanisms to subvert the other. Moreover, the connection between TA systems and Abi is an interesting one; it is a link between systems related to general stress and a strategy to counter a very specific predator.

A recent bioinformatics search was conducted which showed the presence of Type III TA loci across a diverse range of bacteria [76]. The structure and function of a ToxIN homologue from *Bacillus thuringiensis* has also been very recently published [77], but while the structure is very similar to that of ToxIN from *Pectobacterium*, thus far there has been no evidence that it confers an abortive infection phenotype. A general link between Type III TA systems and abortive infection is therefore certainly not clear-cut, but further structural and biochemical studies on other related systems should yield more information on the possible molecular connections. This understanding could enable rational engineering of bacteria that are able to resist a greater spectrum of phages, which would be useful in some industrial fermentations. For phage therapy, it would allow for the design of phages that can specifically bypass Abi systems that may be present in the bacterial pathogens they target. And finally, in medicine, small molecules that activate toxin-antitoxin systems or abortive infection pathways could be utilized as future antibacterial agents against pathogenic bacteria. Abortive infection is a fascinating mechanism; the proteins and pathways bacteria employ to prevent phage infection can be adopted for biotechnological, industrial, and medical applications.

References

[1] Brussow H, Canchaya C, Hardt WD (2004) Phages and the evolution of bacterial pathogens: from genomic rearrangements to lysogenic conversion. *Microbiol Mol Biol R 68*: 560-602.

[2] Waldor MK, Friedman DI (2005) Phage regulatory circuits and virulence gene expression. *Curr Opin Microbiol 8*: 459-465.

[3] Groman NB (1953) The Relation of Bacteriophage to the Change of Corynebacterium diphtheriae from Avirulence to Virulence. *Science 117*: 297-299.

[4] Groman NB (1955) Evidence for the active role of bacteriophage in the conversion of nontoxigenic Corynebacterium diphtheriae to toxin production. *J Bacteriol 69*: 9-15.

[5] Chibani-Chennoufi S, Bruttin A, Dillmann ML, Brussow H (2004) Phage-host interaction: an ecological perspective. *J Bacteriol 186*: 3677-3686.

[6] Hendrix RW (2003) Bacteriophage genomics. *Curr Opin Microbiol 6*:506-511.

[7] Wommack KE, Colwell RR (2000) Virioplankton: viruses in aquatic ecosystems. *Microbiol Mol Biol R 64*: 69-114.

[8] Lima-Mendez G, Van Helden J, Toussaint A, Leplae R (2008) Prophinder: a computational tool for prophage prediction in prokaryotic genomes. *Bioinformatics 24*: 863-865.

[9] Comeau AM, Krisch HM (2005) War is peace – dispatches from the bacterial and phage killing fields. *Curr Opin Microbiol 8*: 488-494.
[10] Weinbauer MG (2004) Ecology of prokaryotic viruses. *FEMS Microbiol Rev 28*: 127-181.
[11] Sulakvelidze A, Alavidze Z, Morris JG, Jr. (2001) Bacteriophage therapy. *Antimicrob Agents Ch 45*: 649-659.
[12] Levin BR, Bull JJ (2004) Population and evolutionary dynamics of phage therapy. *Nat Rev Microbiol 2*: 166-173.
[13] Chopin MC, Chopin A, Bidnenko E (2005) Phage abortive infection in lactococci: variations on a theme. *Curr Opin Microbiol 8*: 473-479.
[14] Forde A, Fitzgerald GF (1999) Bacteriophage defence systems in lactic acid bacteria. *A Van Leeuw J Microb 76*: 89-113.
[15] Kutter E, Sulakvelidze A (2004) *Bacteriophages: Biology and Applications* (CRC Press).
[16] Heller KJ (1992) Molecular interaction between bacteriophage and the gram-negative cell envelope. *Arch Microbiol 158*: 235-248.
[17] Manning PA, Puspurs A, Reeves P (1976) Outer membrane of *Escherichia coli* K-12: isolation of mutants with altered protein 3A by using host range mutants of bacteriophage K3. *J Bacteriol 127*: 1080-1084.
[18] Morona R, Henning U (1984) Host range mutants of bacteriophage Ox2 can use two different outer membrane proteins of *Escherichia coli* K-12 as receptors. *J Bacteriol 159*: 579-582.
[19] Drexler K, Dannull J, Hindennach I, Mutschler B, Henning U (1991) Single mutations in a gene for a tail fiber component of an *Escherichia coli* phage can cause an extension from a protein to a carbohydrate as a receptor. *J Mol Biol 219*: 655-663.
[20] Tomas JM, Benedi VJ, Ciurana B, Jofre J (1986) Role of capsule and O antigen in resistance of *Klebsiella pneumoniae* to serum bactericidal activity. *Infect Immun 54*: 85-89.
[21] Hancock RE, Reeves P (1976) Lipopolysaccharide-deficient, bacteriophage-resistant mutants of *Escherichia coli* K-12. *J Bacteriol 127*: 98-108.
[22] Boman HG, Monner DA (1975) Blocking of bacteriophages phi W and phi 5 with lipopolysaccharides from *Escherichia coli* K-12 mutants. *J Bacteriol 121*: 465-470.
[23] Mizoguchi K, Morita M, Fischer CR, Yoichi M, Tanji Y, Unno H (2003) Coevolution of bacteriophage PP01 and *Escherichia coli* O157:H7 in continuous culture. *Appl Environ Microb 69*: 170-176.
[24] Meyer JR, Dobias DT, Weitz JS, Barrick JE, Quick RT, Lenski RE (2012) Repeatability and contingency in the evolution of a key innovation in phage lambda. *Science 335*: 428-432.
[25] Tanji Y, Hattori K, Suzuki K, Miyanaga K (2008) Spontaneous deletion of a 209-kilobase-pair fragment from the *Escherichia coli* genome occurs with acquisition of resistance to an assortment of infectious phages. *Appl Environ Microb 74*: 4256-4263.
[26] Bohannan B, Travisano M, Lenski R (1999) Epistatic Interactions Can Lower the Cost of Resistance to Multiple Consumers. *Evolution 53*: 292-295.
[27] Petty NK, Evans TJ, Fineran PC, Salmond GP (2007) Biotechnological exploitation of bacteriophage research. *Trends Biotechnol 25*: 7-15.

[28] Tuncer Y, Akçelik M (2002) A protein which masks galactose receptor mediated phage susceptibility in *Lactococcus lactis* subsp. *lactis* MPL56. *Int J Food Sci Technol 37*: 139-144.

[29] Nordstrom K, Forsgren A (1974) Effect of protein A on adsorption of bacteriophages to *Staphylococcus aureus*. *J Virol 14*: 198-202.

[30] Sutherland IW (1995) Polysaccharide lyases. *FEMS Microbiol Rev 16*: 323-347.

[31] Labrie SJ, Samson JE, Moineau S (2010) Bacteriophage resistance mechanisms. *Nat Rev Microbiol 8*: 317-327.

[32] Stummeyer K, Schwarzer D, Claus H, Vogel U, Gerardy-Schahn R, Mühlenhoff M (2006) Evolution of bacteriophages infecting encapsulated bacteria: lessons from *Escherichia coli* K1-specific phages. *Mol Microbiol 60*: 1123-1135.

[33] Drexler K, Riede I, Montag D, Eschbach ML, Henning U (1989) Receptor specificity of the Escherichia coli T-even type phage Ox2. Mutational alterations in host range mutants. *J Mol Biol 207*: 797-803.

[34] Akçelik M, Sanlibaba P (2000) Phage resistance in *Lactococcus lactis* subsp. *lactis* strains isolated from traditional fermented milk products in Turkey. *Int J Food Sci Technol 35*: 473-481.

[35] Hill C (1993) Bacteriophage and bacteriophage resistance in lactic acid bacteria. *FEMS Microbiol Rev 12*: 87-108.

[36] Kraus J, Geller BL (1998) Membrane receptor for prolate phages is not required for infection of *Lactococcus lactis* by small or large isometric phages. *J Dairy Sci 81*: 2329-2335.

[37] O'Driscoll J, Glynn F, Fitzgerald GF, van Sinderen D (2006) Sequence analysis of the lactococcal plasmid pNP40: a mobile replicon for coping with environmental hazards. *J Bacteriol 188*: 6629-6639.

[38] Watanabe K, Ishibashi K, Nakashima Y, Sakurai T (1984) A phage-resistant mutant of *Lactobacillus casei* which permits phage adsorption but not genome injection. *J Gen Virol 65*: 981-986.

[39] Lu MJ, Henning U (1994) Superinfection exclusion by T-even-type coliphages. *Trends Microbiol* 2(4):137-139.

[40] Lu MJ, Stierhof YD, & Henning U (1993) Location and unusual membrane topology of the immunity protein of the *Escherichia coli* phage T4. *J Virol 67*: 4905-4913.

[41] Roossinck MJ (2011) The good viruses: viral mutualistic symbioses. *Nat Rev Microbiol 9*: 99-108.

[42] Pingoud A, Fuxreiter M, Pingoud V, Wende W (2005) Type II restriction endonucleases: structure and mechanism. *Cell Mol Life Sci 62*: 685-707.

[43] Roberts RJ, Belford M, Bestor T, Bhagwat S, Bickle TA, et al. (2003) A nomenclature for restriction enzymes, DNA methyltransferases, homing endonucleases and their genes. *Nucleic Acids Res 31*: 1805-1812.

[44] Sumby P, Smith MC (2002) Genetics of the phage growth limitation (Pgl) system of *Streptomyces coelicolor* A3. *Mol Microbiol 44*: 489-500.

[45] Kruger DH, Bickle TA (1983) Bacteriophage survival: multiple mechanisms for avoiding the deoxyribonucleic acid restriction systems of their hosts. *Microbiol Rev 47*: 345-360.

[46] Tock MR, Dryden DT (2005) The biology of restriction and anti-restriction. *Curr Opin Microbiol 8*: 466-472.

[47] McGrath S, Seegers JF, Fitzgerald GF, van Sinderen D (1999) Molecular characterization of a phage-encoded resistance system in *Lactococcus lactis*. *Appl Environ Microb 65*: 1891-1899.

[48] Brouns SJ, Jore MM, Lundgren M, Westra ER, Slijkhuis RJ et al. (2008) Small CRISPR RNAs guide antiviral defense in prokaryotes. *Science 321*: 960-964.

[49] Sorek R, Kunin V, Hugenholtz P (2008) CRISPR – a widespread system that provides acquired resistance against phages in bacteria and archaea. *Nat Rev Microbiol 6*: 181-186.

[50] Westra ER, Swarts DC, Staals RH, Jore MM, Brouns SJ, van der Oost J (2012) The CRISPRs, they are a-changin': how prokaryotes generate adaptive immunity. *Annu Rev Genet 46*:311-339.

[51] Barrangou R, Fremaux C, Deveau H, Richards M, Boyaval P et al. (2007) CRISPR provides acquired resistance against viruses in prokaryotes. *Science 315*: 1709-1712.

[52] Bondy-Denomy J, Pawluk A, Maxwell KL, Davidson AR (2013) Bacteriophage genes that inactivate the CRISPR/Cas bacterial immune system. *Nature 493*: 429-432.

[53] Sing WD, Klaenhammer TR (1990) Plasmid-induced abortive infection in *Lactococci*: a review. *J Dairy Sci 73*: 2239-2251.

[54] Blower TR, Evans TJ, Przybilski R, Fineran PC, Salmond GP (2012) Viral evasion of a bacterial suicide system by RNA-based molecular mimicry enables infectious altruism. *PLoS Genet 8*: e1003023.

[55] Gross SR (1954) Abortive infection of a strain of *Escherichia coli* by coliphage T2. *J Bacteriol 68*: 36-42.

[56] Cluzel PJ, Chopin A, Ehrlich SD, Chopin MC (1991) Phage abortive infection mechanism from *Lactococcus lactis* subsp. *lactis*, expression of which is mediated by an Iso-ISS1 element. *Appl Environ Microbiol 57*: 3547-3551.

[57] Rettenmier CW, Hemphill HE (1974) Abortive infection of lysogenic *Bacillus subtilis* 168(SPO2) by bacteriophage phi 1. *J Virol 13*: 870-880.

[58] Smith HS, Pizer LI, Pylkas L, & Lederberg S (1969) Abortive infection of *Shigella dysenteriae* P2 by T2 bacteriophage. *J Virol 4*: 162-168.

[59] Snyder L (1995) Phage-exclusion enzymes: a bonanza of biochemical and cell biology reagents? *Mol Microbiol 15*: 415-420.

[60] Parma DH, Snyder M, Sobolevski S, Nawroz M, Brody E, Gold L (1992) The Rex system of bacteriophage lambda: tolerance and altruistic cell death. *Gene Dev 6*: 497-510.

[61] Molineux IJ (1991) Host-parasite interactions: recent developments in the genetics of abortive phage infections. *New Biol 3*: 230-236.

[62] Kao C, Snyder L (1988) The *lit* gene product which blocks bacteriophage T4 late gene expression is a membrane protein encoded by a cryptic DNA element, e14. *J Bacteriol 170*: 2056-2062.

[63] Yu YT, Snyder L (1994) Translation elongation factor Tu cleaved by a phage-exclusion system. *Proc Natl Acad Sci USA 91*: 802-806.

[64] David M, Borasio GD, Kaufmann G (1982) Bacteriophage T4-induced anticodon-loop nuclease detected in a host strain restrictive to RNA ligase mutants. *Proc Natl Acad Sci USA 79*: 7097-7101.

[65] Penner M, Morad I, Snyder L, Kaufmann G (1995) Phage T4-coded Stp: double-edged effector of coupled DNA and tRNA-restriction systems. *J Mol Biol 249*: 857-868.

[66] Amitsur M, Levitz R, Kaufmann G (1987) Bacteriophage T4 anticodon nuclease, polynucleotide kinase and RNA ligase reprocess the host lysine tRNA. *EMBO J 6*: 2499-2503.

[67] Anba J, Bidnenko E, Hillier A, Ehrlich D, Chopin MC (1995) Characterization of the lactococcal *abiD1* gene coding for phage abortive infection. *J Bacteriol 177*: 3818-3823.

[68] Haaber J, Samson JE, Labrie SJ, Campanacci V, Cambillau C, Moineau S, Hammer K. (2010) Lactococcal abortive infection protein AbiV interacts directly with the phage protein SaV and prevents translation of phage proteins. *Appl Environ Microbiol 76*: 7085-7092.

[69] Gerdes K, Christensen SK, Lobner-Olesen A (2005) Prokaryotic toxin-antitoxin stress response loci. *Nat Rev Microbiol 3*: 371-382.

[70] Pecota DC, Wood TK (1996) Exclusion of T4 phage by the *hok/sok* killer locus from plasmid R1. *J Bacteriol 178*: 2044-2050.

[71] Hazan R, Engelberg-Kulka H (2004) Escherichia coli *mazEF*-mediated cell death as a defense mechanism that inhibits the spread of phage P1. *Mol Genet Genomics 272*: 227-234.

[72] Otsuka Y, Yonesaki T (2012) Dmd of bacteriophage T4 functions as an antitoxin against Escherichia coli LsoA and RnlA toxins. *Mol Microbiol 83*: 669-681.

[73] Fineran PC, Blower TR, Foulds IJ, Humphreys DP, Lilley KS, Salmond GP (2009) The phage abortive infection system, ToxIN, functions as a protein-RNA toxin-antitoxin pair. *Proc Natl Acad Sci USA 106*: 894-899.

[74] Blower TR, Fineran PC, Johnson MJ, Toth IK, Humphreys DP, Salmond GP (2009) Mutagenesis and functional characterization of the RNA and protein components of the toxIN abortive infection and toxin-antitoxin locus of Erwinia. *J Bacteriol 191*: 6029-6039.

[75] Blower TR, Pei XY, Short FL, Fineran PC, Humphreys DP, Luisi BF, Salmond GP (2011) A processed noncoding RNA regulates an altruistic bacterial antiviral system. *Nat Struct Mol Biol 18*: 185-190.

[76] Blower TR, Short FL, Rao F, Mizuguchi K, Pei XY, Fineran PC, Luisi BF, Salmond GP (2012) Identification and classification of bacterial Type III toxin-antitoxin systems encoded in chromosomal and plasmid genomes. *Nucleic Acids Res 40*: 6158-6173.

[77] Short FL, Pei XY, Blower TR, Ong SL, Fineran PC, Luisi BF, Salmond GP (2012) Selectivity and self-assembly in the control of a bacterial toxin by an antitoxic noncoding RNA pseudoknot. *Proc Natl Acad Sci USA 110*: 241-249.

In: Bacteriophages
Editors: Clark Denton and Richard J. Crosby
ISBN: 978-1-62618-513-5

Chapter 4

Bacteriophages as Biocontrol Tools of Pathogens in Meat and Dairy Products

David Tomat[1,*], Diego Mercanti[2], Claudia Balague[1] and Andrea Quiberoni[2]
[1]Cátedra de Bacteriología, Área Microbiología,
Facultad de Ciencias Bioquímicas y Farmacéuticas,
Universidad Nacional de Rosario, Rosario, Argentina
[2]Instituto de Lactología Industrial (UNL-CONICET),
Facultad de Ingeniería Química,
Universidad Nacional del Litoral, Santiago del Estero,
Santa Fe, Argentina

Abstract

Biocontrol by bacteriophages in meat and dairy products was developed in the last years and their recent "generally recognized as safe" (GRAS) designation and approval as food additives (FDA, 2006) has opened the discussion about "edible virus". The phage treatment can be applied at different stages in the food chain production, from the live animal to the manufactured product. In this chapter we focus on their application during milk and raw meat processing prior to consumption in order to extend the shelf life of these perishable foods. Studies using bacteriophages as biocontrol tools in meat products have focused particularly on *Escherichia coli* O157:H7, but other pathogens such as *Campylobacter jejuni*, *Listeria monocytogenes* and *Salmonella enterica* have also received attention. Even though the major concern in dairy production is preventing phage infection of dairy starter cultures, the aim of this chapter is to address the use of bacteriophages for biocontrol of pathogens like *Staphylococcus aureus*, *Escherichia coli*, *Enterobacter sakazakii*, *Listeria monocytogenes*, *Mycobacterium smegmatis* and *Salmonella enteritidis*. On the other hand, a section will be dedicated to revise the biocontrol of Shiga-toxigenic *Escherichia coli* (STEC) pathogens in meat and dairy foods. STEC is the etiologic agent of hemolytic uremic syndrome (HUS), a disease endemic in countries like Argentine, which has the highest incidence worldwide with

[*] E-mail address: dtomat@fbioyf.unr.edu.ar; Tel.: +54 341 4352949 ; fax: +54 341 4804592.

13.9 cases per 100,000 children under 5 years of age. In spite of the advantages of the use of bacteriophages in food safety, a drawback is the emergence of bacteriophage insensitive mutants (BIMs). Nevertheless, low frequency of emergence, phenotype reversions and variable stability of BIMs were reported to date. Furthermore, phage cocktails can be used to counteract the effect of BIMs occurrence.

Introduction

Foodborne pathogens present in meat and dairy products are responsible for major outbreaks worldwide (Mead et al. 1999; DeWaal et al. 2009; EFSA 2010; Farrokh et al. 2012). In the meat chain production, a properly performed chore, Good Manufacturing Practice (GMP) and a correct packaging aid to ensure food safety. However, contamination during slaughter, insufficient cooking and cross contamination make foodborne pathogens such as *E. coli* O157:H7 a health threat.

In dairy products, the major concern is focused on *Staphylococcus aureus* (Garcia et al. 2007; Obeso et al. 2010; Bueno et al. 2012) and STEC (Farrokh et al. 2012). During the production process, dairy products can be affected by several risk factors difficult to control: use of raw milk in nonindustrial products, contamination of feedstock during the milking process on the farm, or post-processing contamination (after pasteurization) from the industry to the retail market and consumer's home.

The therapeutic potential of phages has been explored since they were discovered by Félix d`Hérelle in 1917. Nevertheless, and mainly due to poor understanding of basic phage biology, lack of quality control, and absence of properly controlled studies, the use of phages as therapeutic agents decreased, tendency influenced also by the discovery of antibiotics (Petty et al. 2007).

Bacteriophages are the most abundant self-replicating entities in our environment and are present in high concentrations in foods and water (Sulakvelidze and Barrow 2005). High numbers are being routinely consumed via our food: more than 10^8 viable phages per gram are often identified on fresh and processed meat and meat products (Kennedy and Bitton 1987).

As they are natural enemies of bacteria, they have great potential to be used as control agents of foodborne pathogens, especially taking into account the increasing incidence of bacterial resistance to antibiotics. On the other hand, phages can be also used, as an environmentally friendly alternative to toxic chemical compounds, for the decontamination of hard surfaces. In this chapter, the advantages, limitations and disadvantages of using phages as innovative biocontrol tools to improve food safety are reviewed.

1. Biocontrol in Meat

Several strategies are currently applied to preserve perishable refrigerated foods and extend their shelf-life. However, physical processes and chemical compounds (preservatives) used for this purpose may alter meat organoleptic properties. Although bacteriophages represent a novel approach, there are no reports of their industrial use to improve safety, even if this "new, ecological and specific" technology may be cheaper than "old" technologies, since phages can be isolated from the environment and are self-replicating entities. On the other hand, their inclusion into a meat product can be seen as a less aggressive approach.

Numerous studies have been conducted to evaluate reduction or elimination of viable bacterial pathogens onto meat product surface using phages. O'Flynn and coworkers (2004) isolated *E. coli* O157:H7-specific phages from fecal matter, and seven out of nine meat samples were completely free of *E. coli* O157:H7 after enrichment, whereas the two positive samples yielded counts of less than 10 CFU/mL.

Moreover, significant reductions were found on ground beef contaminated with *E. coli* O157:H7. The commercial preparation EcoShield™ (formerly ECP-100; a blend of three individual phages that provides broad protection against pathogenic strains of *Escherichia coli* O157:H7 developed by Intralytix, Inc, USA) was able to reduce the concentration of *E. coli* O157:H7 viable cells by 94.5% during storage at 10°C for 24 h (Abuladze et al. 2008). Foodborne pathogens like *Campylobacter jejuni* are naturally found in the intestine of cattle, though the route of contamination of commercial meat products is not clear (Vashin and Stoyanchev 2005).

C. jejuni phage Cj6 was evaluated for its ability to reduce viable cells from raw and cooked beef at two temperatures (5°C and 24°C) and at low and high either host density (10^2 and 10^4 cells/cm^2) or multiplicity of infection (MOI; 10^1 and 10^4) values. Larger reductions were observed at high host density and high MOI values (Bigwood et al. 2008).

On the other hand, although greater reductions were observed at 24°C during short incubation times, influence of temperature was not clear, since host decreased at 24°C even in the absence of phage. Bigwood et al. (2008) also described the biocontrol of *Salmonella enterica* by phage P7.

In their study, the highest reductions were obtained in the same conditions as for biocontrol of *Campylobacter jejuni*: high host density (10^4 cells/cm^2) and high MOI (10^4), with cell reductions ranging from 1.8 to almost 6 $\log_{10}$ units on raw meat, from 2.0 to 4.5 $\log_{10}$ units on cooked meat at 24°C, and slightly lower though still significant cell reductions at 5°C. On the contrary, when low host density (10^2 cells/cm^2) and high MOI (10^4) were used, reductions were significantly lower at 5°C than at 24°C, obtaining 1.2 $\log_{10}$ reductions at best.

In addition, it is worth mentioning that the higher the host density, the higher the reductions obtained, regardless of the temperature tested. This might be due to the presence of more cells available to kill or to a higher number of infected cells, thus contributing to further phage propagation and subsequent cell lysis. Regarding the control of *Listeria monocytogenes* in meat, several authors found contrasting results. Dykes and Moorehead (2002) showed that the combination of a *L. monocytogenes* phage at 3×10^3 PFU/mL (MOI= 10^{-3}) and nisin (5×10^3 IU/mL) provided an antimicrobial effect against *L. monocytogenes* in broth at 7°C or 30°C, but not in buffer or in raw beef at 4°C. Furthermore, listeriophage alone had no effect in broth under the conditions studied.

Also, a fast but small (1 $\log_{10}$) reduction of *L. monocytogenes* was observed on slices of cooked ham when phages were added at a MOI of 5×10^4, though surviving host cells grew, similarly to the phage-free controls, reaching 10^7 CFU/g after 14 days at 10°C (Holck and Berg 2009). On the other hand, Guenther et al. (2009) reported reductions on sausages of up to 4.2 and 4.5 $\log_{10}$ after 6 and 13 days of storage at 6°C, respectively. However, they concluded that phage concentration should not be lower than 10^8 PFU per g or cm^2 at the time of application to achieve effective results.

Table 1. Reported phage-based biocontrol in meat products

Bacterial target	Biopreserved food	Phage/derived product	References
Escherichia coli O157:H7	Meat	Phages e11/2, pp01 and e4/1c	O'Flynn et al. 2004
	Beef	EcoShield™ (ECML-4 ECML-117 and ECML-134)	Abuladze et al. 2008
Campylobacter jejuni	Cooked and raw meat	Phage Cj6	Bigwood et al. 2008
Salmonella enterica		Phage P7	
Listeria monocytogenes	Raw beef	Phage LH7	Dykes and Moorehead 2002
	Cooked ham	Listex™ P100	Holck and Berg 2009
	Sausages	Phages A511 and P100	Guenther et al. 2009

Phages were in most cases successful to control foodborne pathogens on meat, though several requirements must be fulfilled in all food systems, either solid or liquid:

1) Phages must be lytic: i.e. should not be integrated into the bacterial genome to replicate; they must multiply and lyse the host instead.
2) Non-transducing: i.e. phages must not incorporate bacterial genes in their genomes, given that a new phage with foreign genes may infect sensitive bacterial cells, thus spreading pathogenicity factors.
3) Lack of any virulence trait: phages used for this purpose should not possess virulence factors in their genomes to ensure that no gene transfer of these resistance/virulence features occurs.
4) High MOI: studies largely agree in using high MOI values to obtain a large bacterial reduction. This condition is to guarantee that at least one phage reached every target bacterium, as phages are not freely diffusible in a solid matrix such as meat.

Even when phages meet these conditions in order to be used for biocontrol of pathogens in food, more research, empirical data and a better understanding of phage-host interactions that take place at meat matrix is still needed.

2. Biocontrol in Dairy Products

Likewise meat industry, dairies use physical and chemical methods of conservation with the same undesirable effects. Another method used in meat is the addition of protective cultures (Holck and Berg 2009). However, dairy starter cultures are made up of lactic acid bacteria (LAB) belonging to genera such as *Streptococcus*, *Lactobacillus*, *Lactococcus* and

Leuconostoc, which provide this protective function by themselves through the fermentation process. However, post-pasteurization contamination of feedstock can lead to reductions in LAB efficiency, therefore modifying the quality and taste of end-products and causing economic losses for producers as well as food poisoning in consumers. Furthermore, some dairy products are made from raw milk, another major concern regarding food safety.

Special attention must be paid to *Staphylococcus aureus*, since it is one of the pathogenic bacteria considered as a major threat for public health. *S. aureus* is responsible for foodborne outbreaks associated with milk and dairy products (de Buyser et al. 2001; Tirado and Schmidt 2000; EFSA and ECDC 2011). Staphylococcal food poisoning is due to staphylococcal heat-stable enterotoxins preformed in food (Le Loir et al. 2003) and bovine mastitis caused by *S. aureus*, the most important source of milk contamination with this pathogen (Katsuda et al. 2005). However, post-pasteurization contamination by food handlers has also been reported (Waldvogel 2001). Challenge tests were performed in ultra-high-temperature (UHT) whole milk at 37°C with two lytic phages (MOI=$1x10^2$) that were able to eliminate $3x10^6$ CFU/mL of *S. aureus* after 2 h of incubation (Garcia et al. 2007). In addition, the same authors investigated the biocontrol effect of phages during curdling and found that in acid curd manufacture at 25°C (MOI = $2.5x10^2$), complete clearance of $1x10^6$ CFU/mL was achieved at 4 h, whereas in enzymatic curd manufacture at 30°C (MOI = $3.5x10^2$) complete *S. aureus* inactivation was observed within 1 h of incubation. During challenge assays in commercial pasteurized whole milk, Obeso et al. (2010) found that, at high phage concentration ($\geq$ 6 log_{10} PFU/mL), the phage mixture used completely prevented *S. aureus* growth after 2 h of incubation at high initial level of bacterial contamination ($\geq$ 6 log_{10} CFU/mL), while at lower bacterial contamination level ($\leq$ 4 log_{10} CFU/mL), a notable delay in the inhibitory effect of a low phage concentration (2 to 3 log_{10} PFU/mL) was observed. In addition, they inferred that the closer to the optimal bacterial growth temperature, the greater the phage inhibitory activity. Cheese-making experiments were conducted in 12 L vats (pilot plant scale) to evaluate the same two lytic phages previously tested in UHT whole milk by Garcia (2007), but this time during curdling of fresh and hard cheeses. In fresh cheese, a *S. aureus* initial inoculum of $1.7x10^6$ CFU/mL treated with $1x10^6$ PFU/mL of phages (MOI of about 1) resulted in a 3.83 log_{10} reduction at 3 h and below the detection limit at 6 h, though at this point viable cells were undetected also in control samples, suggesting that other factors are involved in killing *S. aureus* population.

Similar results were found in hard cheese, where the starting inoculum of *S. aureus* ($8x10^5$ CFU/mL) was significantly reduced after treatment with $5x10^6$ PFU/mL of the phage cocktail during curdling. On the other hand, no re-growth during 14 days of storage at 4°C was observed for fresh cheese, whereas for the hard cheese complete inactivation was never reached and 1.24 log_{10} of cells remained still viable at the end of ripening (30 days) (Bueno et al. 2012).

Although phage treatments were effective in this study, for both fresh and hard cheeses, viable cell decreases due to additional factors such as pH, bacteriocins and H_2O_2 partially contributed to obtain the large *S. aureus* reductions observed.

Cattle are a reservoir of *E. coli* and faecal contamination can occur during milking as well as post-pasteurization. However, reports related to phage biocontrol of *E. coli* in milk are very scarce. Moradpour et al. (2009) evaluated a genetically modified bacteriophage at a MOI of 100 to reduce *E. coli* O157:H7 in milk ($1x10^7$ CFU/mL), and after 120 min they obtain curves of first order kinetics for either growth (without phage) or inactivation (with phage) achieving a large reduction, though bacterial survival was still observed ($1x10^1$ CFU/mL).

A major concern in powdered infant formula is the pathogen *Enterobacter sakazakii* (also known as *Cronobacter* spp.) implicated in outbreaks and causing high mortality and sequels on children. Phages assayed on this pathogen showed bacteriostatic rather than bacteriolytic activity. Kim et al. (2007) evaluated two phages for *E. sakazakii* biocontrol on infant formula (10^2 CFU/mL) at three phage concentrations (10^7, 10^8 and 10^9 PFU/mL) and three temperatures (12°C, 24°C and 37°C). Either at12°C or 37°C, only one phage (ESP732-1) at the highest concentration tested showed bacteriolytic activity and eliminated *E. sakazakii* completely; lower phage concentrations resulted in a bacteriostatic activity followed by a culture re-growth. On the other hand, no re-growth was observed at 24°C for both phages added at 10^8 and 10^9 PFU/mL, thus being the best biocontrol conditions assayed. In addition, samples of infant formula contaminated with *E. sakazakii* and treated with a cocktail of five phages (MOI=10^6) resulted in the prevention of growth of 36 out of 40 tested strains; phages resulted bacteriolytic in this case, though the same cocktail at a MOI of 10^2 failed to prevent outgrowth of the strains (Zuber et al. 2008).

Listeria monocytogenes is an environmental bacterium resistant to different physical, chemical and biological agents (Pearson and Marth 1990; Tasara and Stephan 2006). Although pasteurization is a safe treatment, there exists the risk of later recontamination especially on dairy products such as soft cheese.

Application of listeriophage P100 on an artificial cheese surface model (Ch-easy plates) and during a commercial cheese production process defined optimal conditions; either repeated phage doses at high concentration (0.2 mL of $3x10^9$ PFU/mL, resulting in $6x10^7$ PFU/cm^2) or a single phage dose using a larger volume (1.0 mL of $6x10^8$ PFU/mL) to optimize the phage distribution on the cheese surface completely inactivated *L. monocytogenes*, without further re-growth (Carlton et al. 2005). A P100-related phage (A511) was evaluated by Guenther et al. (2009) for biocontrol in chocolate milk and mozzarella cheese brine artificially contaminated with *L. monocytogenes* ($1x10^3$ CFU/g). Reductions with respect to control samples without phages ranged from 2.3 to 5.0 $\log_{10}$ units after 6 days at 6°C, when using a MOI of $3x10^5$.

Also, an extended experiment over 13 days at 6°C on both food matrices (chocolate milk and mozzarella cheese brine) resulted in a complete eradication of *L. monocytogenes*. However, for assays carried out in chocolate milk at either 20°C or 6°C and lower MOI values ($3x10^3$ and $3x10^4$) for 6 days, re-growth was observed.

Fast-growing acid-fast bacteria such as *Mycobacterium smegmatis* have been isolated from cows with mastitis, though transmission of viable mycobacteria to humans through pasteurized dairy products seems unlikely (Schultze and Brasso 1987).

However, non-pathogenic *M. smegmatis*, with a 3-4 hour generation time and available genomic sequence, results a good general model for the study of *Mycobacterium* spp. Six individual *M. smegmatis* phages (10^5 to 10^9 PFU/mL) and a mix thereof (10^9 PFU/mL) were assessed to determine whether they could inhibit the growth of *M. smegmatis* ($1x10^3$ CFU/mL) in 10% reconstituted skim milk at 37°C over 4 days. When a single phage was used, a bacteriostatic effect was observed at a MOI between 10^2 and 10^5, and evident at 10^6, though in all cases re-growth took place. On the contrary, when phage cocktail was used *M. smegmatis* cells were completely inactivated, concluding that phage titres must be at least of $1x10^9$ PFU/mL in order to exert a complete bactericidal effect (Endersen et al. 2012).

Pathogens like *Salmonella* spp. were found, at low but still worrying frequency (5.9 %: 5 out of 85 raw milk samples), in dairy products elaborated from raw milk (Addis et al. 2011).

Modi et al. (2001) conducted experiments where raw and pasteurized milk (for cheese making) were inoculated with *Salmonella enteritidis* (10^4 CFU/mL) and phage SJ2 at 10^8 PFU/mL. Small reductions of ca. 2 $\log_{10}$ units were observed during cheese making (24 h) in both types of milk, though a decreasing trend and a complete inactivation of *S. enteritidis* were achieved in cheeses made from raw and pasteurized milk, respectively, after 89 days of ripening in the presence of phage.

Previous reports have shown that milk whey proteins could inhibit phage adsorption to the cell surface (Gill, et al. 2006), but pH is the most crucial inactivation factor (Garcia et al. 2007), somewhat explaining the small reductions at the beginning of cheese ripening and *S. enteritidis* decrease/eradication after 89 days, respectively.

Another serovar causing human illness through contaminated milk products is *Salmonella* Typhimurium.

Table 2. Reported phage-based biocontrol in dairy products

Bacterial target	Biopreserved food	Phage/derived product	References
Staphylococcus aureus	Curd	Phages Φ35 + Φ88	Garcia et al. 2007
	Commercial UHT whole milk	Phages ΦH5 + ΦA72	
	Commercial whole milk	Phages phiIPLA35 and phiIPLA88	Obeso et al. 2010
	Fresh and hard type cheese	Phages phiIPLA35 and phiIPLA88	Bueno et al. 2012
Escherichia coli O157:H7	Milk	Genetically modified phage	Moradpour et al. 2009
Enterobacter sakazakii	Powdered infant formula	Phages ESP 1-3 and ESP 732-1	Kim et al. 2007
		Phages F, 23, 81, 82 and 83 cocktail	Zuber et al. 2008
Listeria monocytogenes	Soft cheese	Phage P100	Carlton et al. 2005
	Chocolate milk and mozzarella cheese brine	Phage A511	Guenther et al. 2009
Mycobacterium smegmatis	Reconstituted skim milk	Phages LE1, LE2, LE3, LE4, LE5 and LE6 cocktail	Endersen et al. 2012
Salmonella enteritidis	Cheddar cheese	Phage SJ2	Modi et al. 2001
Salmonella Typhimurium	Chocolate milk	Phage FO1-E2	Guenther et al. 2012

In order to assess the effect of phage FO1-E2 on this pathogen, chocolate milk spiked with *S.* Typhimurium cells ($1x10^3$ CFU/mL) was treated with $3x10^8$ PFU/mL at two temperatures (8ºC and 15ºC) for 6 days. After 1 day at 8ºC, viable cells could not be detected by direct plating but were detected in all phage-treated samples after enrichment, whereas at 15ºC there was a 2 log_{10} reduction after 2 days followed by re-growth during the remaining incubation period (Guenther et al. 2012).

Considering all the information available to date, phages have demonstrated to be useful for the biocontrol of foodborne pathogens present in dairy products.

However, most studies showed that a very high MOI is required to achieve a substantial improvement on food safety; otherwise a bacteriostatic effect is observed, sometimes followed by a further bacterial re-growth. Furthermore, phages can be used as an additional tool in combination with treatments such as bacteriocins and protective cultures to ensure absence of foodborne pathogens in dairy products.

3. STEC Biocontrol in Meat and Dairy Products

Foodborne diseases involving Shiga toxin-producing *E. coli* (STEC) is becoming an increasing problem, especially the recent emerging O157:H7 STEC, a major food pathogen that can cause diarrhea and more severe clinical manifestations including thrombotic thrombocytopenic purpura (TTP), hemorrhagic enterocolitis (HEC) and hemolytic uremic syndrome (HUS) (Griffin et al. 2002; Su and Brandt 1995). Other non-O157 STEC serogroups (O26, O103, O111, O128 and O145) isolated from food, have also been identified as human pathogens and responsible for the incidence of HUS (WHO 1998; Blanco et al. 2003). STEC produce several virulence factors that contribute to their pathogenicity, mainly Shiga toxins (Stx), which inhibit protein synthesis in target cells (Thorpe et al. 2002). It is thought that the Shiga toxin enters the systemic circulation in the gut, resulting in distal organ damage.

Several observations suggest that the risk of serious complications in STEC infection, e.g. HUS, is related to the presence and amount of Shiga toxin produced during infection (Ritchie et al. 2003). A key issue regarding O157:H7 STEC strains is that they have infectious doses as low as 10 CFU. Furthermore, with the increasing incidence of antibiotic resistant STEC, they may be considered as a major foodborne threat.

The main STEC reservoir is cattle and other ruminants which possess STEC in their intestinal flora (Kobayashi et al. 2001), even if carrier animals are asymptomatic. People become infected with these strains primarily through ingestion of contaminated meat -e.g. meat contaminated during slaughter and undercooked- and other food products -e.g. cross contamination or fecal-oral transmission of infected humans or animals (Su and Brandt 1995).

In addition, the manufacture of cheeses with raw milk, or products with slow or poor acidification and unhygienic conditions in slaughterhouses and dairy farms, lead to bacterial pathogen contamination of meat and dairy products.

Within the chain production, Good Hygiene Practice (GHP), Good Manufacture Practice (GMP) and quality and process controls aid to ensure food safety. In addition, strategies should include educational programs to bring awareness of the STEC problem to farmers, food handlers and consumers.

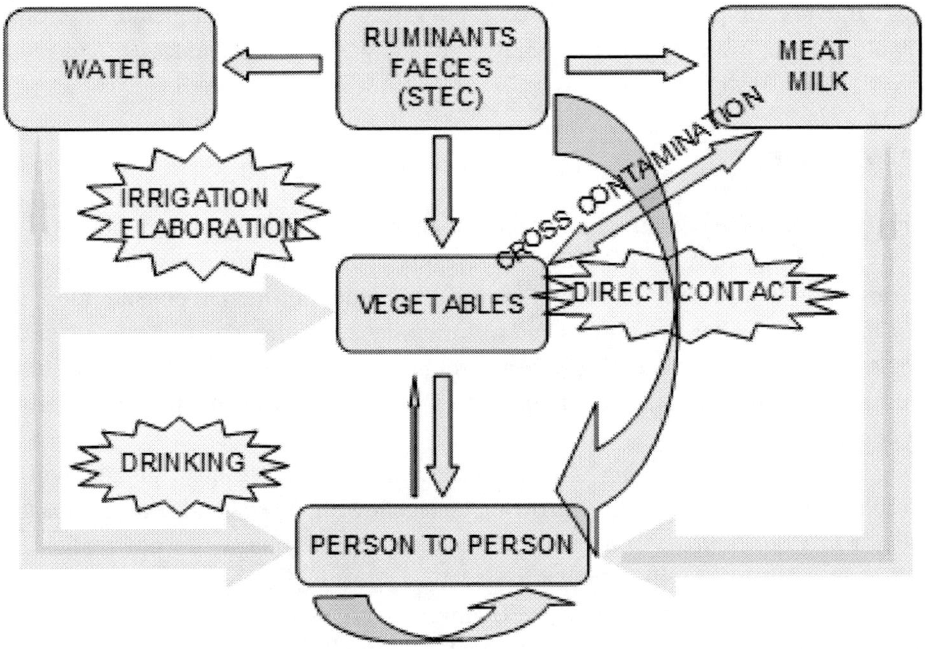

Figure 1. Food contamination routes by STEC.

Regarding STEC biocontrol by bacteriophages, dairy products have received less attention than meat products to date, likely due to the high prevalence of O157:H7 and non-O157 STEC in beef and, above all, minced meat. Although results found by O'Flynn et al. 2004 are encouraging, as most meat samples treated with bacteriophages were freed of the host (see section 1), conditions tested were limited, i.e. they tested only one MOI=10^6 at one temperature (37ºC) and a single sampling point at 1 h of incubation, likewise results obtained by Abuladze et al. 2008 (ca. 10^1, 10ºC and 24h, respectively) though these authors obtained a reduction of only 1.28 $\log_{10}$ (94.5%). Regarding dairy products, even though Moradpour et al. (2009) obtained a 6 $\log_{10}$ reduction, they still have observed *E. coli* O157:H7 survival. Therefore, there is a need to test a more diverse set of conditions so as to obtain more empirical data regarding phage mechanisms of action (lytic or lysis from without, LO) and, in some cases, to obtain greater reductions or complete inactivation. On the other hand, it is mandatory to find optimal biocontrol conditions for each particular food/pathogen/phage system in order to ensure that foods will get STEC-free by using this "new" phage-technology.

4. Bacteriophages: Advantages and Disadvantages as Biocontrol Tools

Phages are everlasting natural entities, whose isolation is easy and less expensive than developing new drugs such as antibiotics. Furthermore, while the latter become ineffective due to bacterial resistance, phages do not, since constantly evolve and can adapt *in situ* to reinfect resistant bacteria, or can be otherwise used as cocktails to minimize resistance.

Although the use of phages to improve food safety is a promising and environmentally friendly alternative, it has, as any technology, advantages and disadvantages.

4.1. Advantages

Phages are highly specific: due to receptor-ligand interactions, phages posses a high host specificity affecting only target bacteria. Besides, they do not cross species or genera boundaries. This is especially important for foods intended for human consumption, since it is unlikely that phages alter the normal microflora balance of the gastrointestinal tract. Bruttin and Brüssow (2005) administered *Escherichia coli* phage T4 orally to human volunteers and observed no adverse effects, concluding that phages employed in the study were safe for human consumption.

Phages are self-perpetuating: in the presence of susceptible host, phages will replicate and subsequently infect other active growing cells, showing a self-amplification response.

Phages are self-limited: in absence of susceptible host phages do not replicate, as they need a host to do that and will only persist as long as host is present.

Phages do not affect eukaryotic cells: there is no evidence to date that these viruses represent a threat to eukaryotic cells (Barrow and Soothill 1997).

Phages do not affect food quality: phages are present and consumed at daily basis in foods and water. In this regard, it is more likely that phages contribute to balance the growth of spoilage microorganisms. Information concerning undesirable organoleptic changes is however scarce, and there is no reports of negative changes produced by addition of bacteriophages to foods (Greer 1986).

Natural, ubiquitous and abundant entities: phages are natural enemies of bacteria as well as the most abundant entities on the planet (ca. 10^{32} phages), distributed among all environments such as soil, water, foods, etc. (Rohwer and Edwards 2002).

Easy to isolate and propagate: phages can be readily isolated from food or water and propagated on their host, usually yielding relatively high titres (10^9 PFU/mL) at very low cost.

4.2. Disadvantages

Consumer acceptance: the main hurdle to overcome in order to use phages to improve food safety may be consumer acceptance. Nevertheless, by providing the consumers with the right information, this new technology would eventually be accepted.

Phage mutation: in general, spontaneous mutations are deleterious to bacteria, though in some cases they can generate transitory advantages while phage is present.

Mutation to temperate phage: a lysogenic conversion may occur where phage DNA is inserted into the bacterial chromosome; in this case cells are no longer sensitive to phage attack. Moreover, this may lead to phage gene expression, generating undesirable characteristics.

Mutation to Bacteriophage Insensitive Mutant (BIM): loss of sensitivity caused by a bacterial mutation may result in selection due to phage infection and elimination of competing sensitive bacteria, leading to a re-growth of the insensitive clone.

Limited understanding of phage-bacteria interactions: phages and bacteria co-evolved so they must co-exist; if a phage kills all the susceptible host cells, it can no longer replicate. Accordingly, phages require a bacterial threshold to replicate (10^3 to 10^5 CFU/mL) and effectively kill target cells (Wiggins and Alexander 1985). Currently, the underlying mechanisms supporting this balance are unknown, and their full understanding would allow obtaining better phages for biocontrol of pathogens.

Transduction: although phages used for this purpose must be strictly lytic, transduction is an issue to be addressed, as what can actually happen in food or after food consumption is not completely known.

Limited host range: a single phage is not able to kill all the pathogenic strains that can potentially be present in food, although using a mixture of different phages can broaden the host range and therefore improve the potential of the treatment (Barrow and Soothill 1997).

Limited diffusion in solid food: phages are not freely diffusible in solid foods such as meat, due to the intricate network of proteins and fibers. Nevertheless, all the authors reviewed in this chapter agree that this disadvantage can be bypassed by adding large amounts of phage (at least 10^8 PFU/mL).

5. Bacteriophage Insensitive Mutants

Currently, the emergence of bacteriophage insensitive mutants (BIMs) during experiments on dairy and meat products is likely to be the most important impediment to develop successful protocols of phage biocontrol in food. BIMs generally arise from point mutations in genes encoding receptor molecules on the bacterial cell surface (Forde and Fitzgerald, 1999). Factors contributing to phage resistance include alteration in the length of lipopolysaccharide (LPS), regulation of its production, and complete loss of receptors. Nevertheless, these alterations give the bacteria only a temporary advantage in the presence of phage and might be also detrimental. Most authors cited in this review had to deal with this issue; some of them deepened BIM analysis to determine occurrence frequencies, reversion and stability to elucidate their implication during phage biocontrol in foods.

Regarding the evaluation of BIM emergence and concerning specifically *E. sakazakii*, 18 out of 20 colonies randomly selected were resistant to the phage, being responsible for re-growth observed in brain heart infusion (BHI) broth (Kim et al. 2007). Studies performed on *S. aureus* determined reversion frequencies of $2.5 \pm 0.12 \times 10^{-4}$ and $1.3 \pm 0.24 \times 10^{-6}$ for temperate and lytic phages, respectively. The highest reversion frequency was observed for temperate phages, because of their inability for super infection (Garcia et al. 2007). Kudva et al. (1999) observed growth of *E. coli* O157:H7 in cultures at 8, 24 and 48 h after phage infection; surviving cells were still sensitive when challenging with phages, but after 5 days became resistant to the three phages evaluated. In addition, O'Flynn et al. (2004) further analyzed *E. coli* O157:H7 BIMs arising from challenging at 37°C and found morphological changes in shape from rod to small cocci; although the growth rates of BIMs were similar to that of the parental strain, the turbidity achieved by the former was always lower. Moreover, *E. coli* O157:H7 BIMs derived from *Myoviridae* and *Siphoviridae* phages arose at frequencies between $1.2 \pm 0.49 \times 10^{-6}$ and $1.9 \pm 0.57 \times 10^{-6}$, and $3.3 \pm 1.1 \times 10^{-4}$, respectively (O'Flynn's,

2004). Furthermore, after 50 generations all BIMs phenotype reverted when challenged with each phage.

BIMs can also be found in dairy products after phage treatment. Endersen et al. (2012) evaluated 10 presumptive BIMs of mycobacteria isolated from challenging in reconstituted skimmed milk (RSM); when BIMs were propagated in BHI broth resulted in growth despite phage presence, although it could not be confirmed due to BIMs aggregation in liquid cultures.

Similar results were obtained in chocolate milk after 6 days at 15ºC, (where 10 *Salmonella* Typhimurium BIMs randomly selected were insensitive to FO1-E2 phage (Guenther et al. 2012). On the other hand, more than 30 presumptive *Listeria monocytogenes* BIMs purified from Ch-easy plates and cheese (treated with low doses of phage P100) were found to be sensitive after subsequent challenging with phage P100 (Carlton et al. 2005).

Contrary to the findings in dairy products, no bacterial resistance (BIMs) was detected after phage treatment of meat. Regarding to *Salmonella* BIMs, none of 27 isolates obtained from phage-treated meat samples was susceptible to subsequent infection by phage P7 (Bigwood et al. 2008). Moreover, other authors were unable to find *E. coli* O157:H7 and *Listeria monocytogenes* BIMs. Abuladze et al. (2008) tested five randomly selected *E. coli* O157:H7 colonies isolated at the end of experiments (ground beef treated with ECP-100 phages at 10ºC) and all potential BIM assessed were sensitive to lysis by ECP-100. Concerning *Listeria*, ten presumptive BIMs isolated by Guenther et al. (2009) from phage treated hot dogs after 6 and 13 days at 6ºC were checked for phage A511 resistance by spot test. Although *Listeria* re-growth was observed in phage treated hot dogs, all the isolates were sensitive to phage A511.

Although bacterial re-growth was observed in most studies conducted in solid food, BIMs emergence seems lower or even null in this kind of matrices. This may be due to poor phage diffusion or because bacteria remain hidden in the protein network of solid foods such as meat.

Therefore, most bacterial colonies isolated after phage treatment (presumptive BIMs) on solid food products resulted phage sensitive, likely because they had never been in contact, thus escaping phage treatment. In this regard, most authors suggest that phage treatments applied on solid food need improvement, by adding for example higher loads of phages.

6. Phage Cocktails for BIMs Emergence Reduction

One of the strategies used to prevent BIMs development is utilizing lytic phage mixtures, otherwise referred as phage cocktails. The rationale supporting this biocontrol strategy is the following: different phages from a given cocktail ($phage^1$, $phage^2$, …, $phage^N$) bind to different receptors (R^1, R^2, …, R^N) on the same bacterial species; mutations generating $phage^{1\text{-}R1}$ BIMs should not affect bacterial sensitivity for $phage^{2\text{-}R2,\ 3\text{-}R3,\ldots N\text{-}RN}$, while $phage^{2\text{-}R2}$ BIMs should still be sensitive to $phage^{1\text{-}R1,\ 3\text{-}R3,\ldots N\text{-}RN}$, and so on.

This would make less likely the survival of BIM arising from different phages. Therefore, the more phages are added to a cocktail, the lower the probability that a resistance event becomes evident. Taking into account this behavior, cocktails of up to six phages were developed for use in foods.

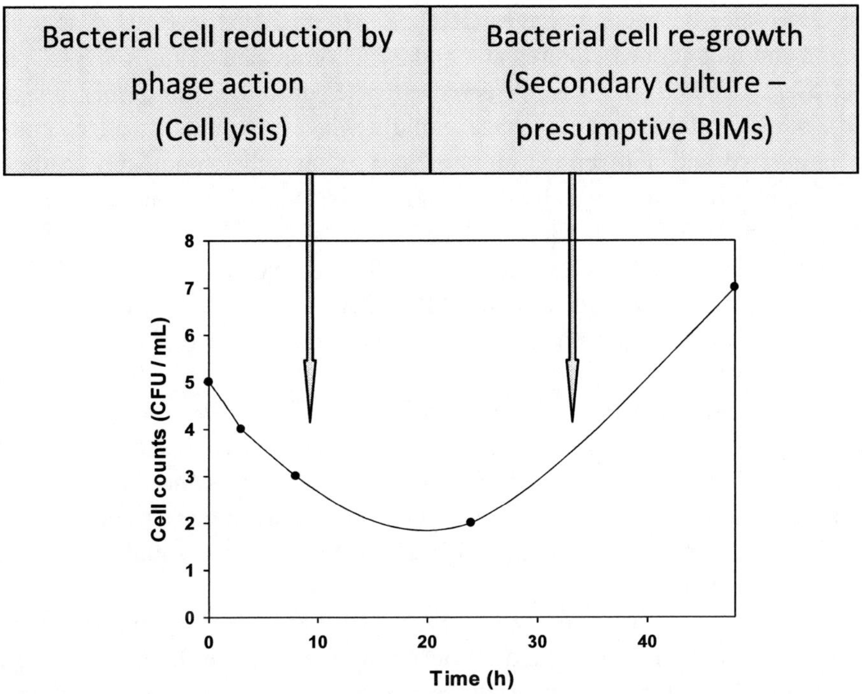

Figure 2. Curve of theoretical reduction of host cells by phage action and subsequent secondary culture development due to bacterial re-growth (presumptive BIMs development).

Table 3. Summary of the efficacy of treatments with phage cocktails

Phage cocktail	Cocktail maximun efficacy (optimal conditions)/ observations	Food/matrix	Reference
Three *E. coli* O157:H7 phages	Complete elimination	Broth	Kudva et al. 1999
Two *S. aureus* lytic phages	Complete elimination	Curd manufacture UHT whole milk	Garcia et al. 2007
Two *S. aureus* temperate phages	3.5 log reduction/no BIMs development	UHT whole milk	
Six mycophages	Complete elimination	RSM	Endersen et al. 2012
EcoShield™ (Three *E. coli* O157:H7 phages)	1.28 log reduction/ no BIMs development	Ground beef	Abuladze et al. 2008
Three *E. coli* O157:H7 phages	Complete elimination in 7 out of 9 samples/no BIMs assessed	Meat	O'Flynn et al. 2004

Three O157-specific bacteriophages (KH1, KH4 and KH5) were used as a cocktail. In either aerated and non-aerated cultures at 4 and 37ºC, or aerated cultures at 37ºC, a complete *E. coli* O157:H7 cells lysis was observed after 5 days by using this phage cocktail (MOI=10^3), whereas no individual phage could achieve a complete cell clearance (BIMs emerged) in aerated and non-aerated cultures at 37ºC and the same MOI. Although phage cocktail was more efficient than individual phages, it still gave place to re-growth of cells in non-aerated cultures at 37ºC (Kudva et al. 1999).

In relation to dairy products, Garcia et al. (2007) compared the biocontrol ability of two phage cocktails against *S. aureus*, at equal MOI (100) and 37ºC, on UHT whole milk. The first one consisted of two lytic phages (Φ88 + Φ35) and the second one of two temperate phages (ΦH5 + ΦA72).

Temperate phage cocktail significantly reduced cell numbers (3.5 $\log_{10}$) after 8 h of incubation, though lytic phage cocktail achieved a faster (2 h) and complete pathogen elimination, therefore showing greater effectiveness and at the same time preventing BIMs emergence. In addition, when lytic phage cocktail was assayed during acid (MOI=250) and enzymatic (MOI=350) curd manufacture, viable *S. aureus* cells were undetected after 4 h and 1 h of incubation, respectively.

Endersen et al. (2012) observed and analyzed BIMs isolated from individual mycophage challenges at $1x10^9$ PFU/mL in RSM, after observing re-growth at 96 h. However, when they used a cocktail of 6 mycophages at the same titre ($1x10^9$ PFU/mL) and in the same experimental conditions (10^3 CFU/mL, RSM at 37ºC), mycobacterial cells were completely inactivated after 96 h, demonstrating the efficacy and usefulness of phage cocktails. Low cocktail MOIs gave place to host re-growth (non-confirmed, presumptive BIMs), remarking the importance of using high phage titres as well.

Meat-borne pathogen *E. coli* O157:H7 treated with a commercial phage cocktail (EcoShield™) was able to survive at $3.4x10^3$ CFU/g (5% of the initial inoculum) on ground beef, but all the surviving bacterial colonies examined showed no resistance to EcoShield™ (Abuladze et al. 2008).

In another study, two out of nine meat samples were positive for *E. coli* O157:H7 after cocktail treatment, but resistance was not assessed (O'Flynn et al. 2004). Listex™ P100, a phage product developed by the company EBI Food Safety (www.ebifoodsafety.com), recently approved by the FDA as a food biopreservative and granted as GRAS (Generally Recognised As Safe), was tested on cooked ham. *L. monocytogenes* grew in Listex™ P100 treated samples, as well as in the phage-free control sample, though no BIMs emergence was observed (Holck and Berg 2009).

Intralytix, Inc. also received FDA approval for using a bacteriophage-based preparation, namely ListShield (six naturally occurring bacteriophages isolated from the environment), as a food additive for biocontrol of *Listeria monocytogenes* that can be used on ready-to-eat (RTE) foods. Currently, several companies situated in the USA are involved in phage cocktails development, e.g., Exponential Biotherapies (www.expobio.com); Gangagen (www.gangagen.com); Intralytix and Omnilytics (www.phage.com). In addition, six companies are developing this technology in other countries such as Phage Biotech (Israel) (www.phage-biotech.com); Hexal Genentech (Germany) (www.hexal-gentech.com); Novolytics (UK) (www.novolytics.co.uk); Biophage (Canada) (www.biophagepharma.net); Biopharm Pharmaceuticals (Georgia) (www.biopharmservices.com); and EBI Food Safety (Netherlands).

Conclusion

When using phages as biocontrol tools in meat products, either contrasting (e.g., for *Listeria monocytogenes*) or encouraging (e.g., for *E. coli* O157:H7) results were obtained, though there are several common requirements that must be fulfilled by phages in any food system (see section 1). Also, more empirical data of phage-host interactions involved at the meat matrix are needed. Regarding dairy products, successful biocontrol was achieved for *S. aureus*, the major threat in milk products. In most dairy assays, however, a very high MOI should be used; otherwise, a bacteriostatic effect is observed followed by a subsequent bacterial re-growth. On the other hand, combination of phages with different approaches such as bacteriocins and/or protective cultures represents a good alternative to control foodborne pathogens on dairy products. Application of chemical and physical treatments is restricted to certain points throughout the food manufacture process. These treatments may alter food organoleptic characteristics and food might also suffer ulterior contaminations. On the contrary, phages may be applied at any moment throughout the production process, and their activity depends upon bacterial cell density and temperature. Therefore, phages can act after a breakdown within the cold chain; if this occurs, the rise of the temperature enhances bacterial metabolism and, in consequence, cell replication, but also phage replication, thus promoting host cell lysis. Superinfection by the same or different phages to avoid pathogen re-growth might be an interesting alternative to the use of phage cocktails, as most authors observed a reduction on bacterial cells population during the first hours post-infection and bacterial re-growth is not due to BIMs emergence.

Beyond what has been previously discussed, the use of phages to improve food safety is a technology developed during the last years, and its application has been delayed for two main reasons; on one hand, more phage biocontrol and viability assays on different foods are needed, testing more realistic food processing conditions. On the other hand, there is still a lack of a regulatory framework in most countries, thus obstructing the application of this procedure.

References

Abuladze, T., Li, M., Menetrez, M. Y., Dean, T., Senecal, A. and Sulakvelidze, A. (2008). Bacteriophages reduce experimental contamination of hard surfaces, tomato, spinach, broccoli, and ground beef by *Escherichia coli* O157:H7. *Applied and Environmental Microbiology,* 74 (20), 6230-6238.

Addis, Z., Kebede, N., Worku, Z., Gezahegn, H., Yirsaw, A. and Kassa, T. (2011). Prevalence and antimicrobial resistance of *salmonella* isolated from lactating cows and in contact humans in dairy farms of Addis Ababa: a cross sectional study. *BMC Infectious Diseases,* 11, 222.

Barrow, P. A. and Soothill, S. (1997). Bacteriophage therapy and prophylaxis: rediscovery and renewed assessment of potential. *Trends in Genetics*, 5, 268-271.

Bigwood, T., Hudson, J. A., Billington, C., Carey-Smith, G. V. and Heinemann, J. A. (2008). Phage inactivation of foodborne pathogens on cooked and raw meat. *Food Microbioloy,* 25, 400–406.

Blanco, J., Blanco, M., Blanco, J. E., Mora, A., González, E. A., Bernárdez, M. I., Alonso, M. P., Coira, A., Rodríguez, A., Rey, J., Alonso, J. M. and Usera, M. A. (2003). Verotoxin-producing *Escherichia coli* in Spain: Prevalence, serotypes, and virulence genes of O157:H7 and non-O157 VTEC in ruminants, raw beef products, and humans. *Experimental Biology and Medicine,* 228, 345–351.

Bruttin, A. and Brüssow, H. (2005). Human volunteers receiving *Escherichia coli* phage T4 orally: a safety test of phage therapy. *Antimicrobial Agents and Chemotherapy*, 49, 2874–2878.

Bueno, E., Garcia, P., Martínez, B. and Rodríguez, A. (2012). Phage inactivation of *Staphylococcus aureus* in fresh and hard-type cheeses. *International Journal of Food Microbiology*, (Article in press).

Carlton, R. M., Noordman, W. H., Biswas, B., de Meester, E. D. and Loessner, M. J. (2005). Bacteriophage P100 for control of *Listeria monocytogenes* in foods: genome sequence, bioinformatic analyses, oral toxicity study, and application. *Regulatory Toxicology and Pharmacology*, 43, 301–312.

De Buyser, M. L., Dufour, B., Maire, M. and Lafarge, V. (2001). Implication of milk and milk products in foodborne diseases in France and in different industrialised countries. *International Journal of Food Microbiology*, 67, 1-17.

DeWaal C. S., Tian, X. A. and Plunkett, D. (2009). Outbreak alert report. analyzing foodborne outbreaks 1998 to 2007. Center for Science in the Public Interest. www.cspinet.org/new/pdf/outbreak_report_2009.pdf.

Dykes, G. A. and Moorehead S. M. (2002). Combined antimicrobial effect of nisin and listeriophage against *L. monocytogenes* in broth but not in buffer or on raw beef. *International Journal of Food Microbiology*, 73, 71-81.

European Food Safety Authority (EFSA) and European Centre for Disease Prevention and Control (ECDC). (2011). The European Union Summary Report on trends and sources of zoonoses, zoonotic agents and food-borne outbreaks in 2009. *EFSA Journal*, 9, 2090, 1-378.

European Food Safety Authority (EFSA). (2010). The Community summary report on trends and sources of zoonoses, zoonotic agents and food-borne outbreaks in the EU. *EFSA Journal*, 8, 1496.

Endersen, L., Coffey, A., Neve, H., McAuliffe, O., Paul Ross, R. and O'Mahony, J. (2012). Isolation and characterisation of six novel mycobacteriophages and investigation of their antimicrobial potential in milk. *International Dairy Journal*, (Article in press).

Farrokh, C., Jordan, K., Auvray, F., Glass, K., Oppegaard, H., Raynaud, S., Thevenot, D., Condron, R., De Reu, K., Govaris, A., Heggum, K., Heyndrickx, M., Hummerjohann, J., Lindsay, D., Miszczycha, S., Moussiegt, S., Verstraete, K. and Cerf, O. (2012). Review of Shiga-toxin-producing *Escherichia coli* (STEC) and their significance in dairy production. *International Journal of Food Microbiology*, (Article in press).

Forde, A. and Fitzgerald, G. F. (1999). Bacteriophage defence systems in lactic acid bacteria. *Antonie Van Leeuwenhoek,* 76, 89–113.

Garcia, P., Madera, C., Martinez, B. and Rodriguez, A. (2007). Biocontrol of *Staphylococcus aureus* in curd manufacturing processes using bacteriophages. *International Dairy Journal*, 17, 1232–1239.

Gill, J. J., Sabour, P.M., Leslie, K.E. and Griffiths, M.W. (2006). Bovine whey proteins inhibit the interaction of *Staphylococcus aureus* and bacteriophage K. *Journal of Applied Microbiology,* 101, 377–386.

Greer, G. G. (1986). Homologous bacteriophage control of *Pseudomonas* growth amd beef spoilage. Journal of Food Protection, 49, 104-109.

Greer, G. (2005). Bacteriophage control of foodborne bacteria. *Journal of Food Protection,* 68 (5), 1102-1111.

Griffin, P. M., Mead, P. S. and Sivapalasingam, S. (2002). *Escherichia coli* O157:H7 and other enterohemorrhagic *Escherichia coli*, p. 627–642. In M. J. Blaser, P. D. Smith, J. I. Ravdin, et al. (ed.), Infections of the gastrointestinal tract, 2nd ed. Lippincott Williams and Wilkins, Philadelphia, Pa.

Guenther, S., Huwyler, D., Richard, S. and Loessner, M. J. (2009). Virulent bacteriophage for efficient biocontrol of *Listeria monocytogenes* in ready-to-eat foods. *Applied and Environmental Microbiology,* 75 (1), 93–100.

Guenther, S., Herzig, O., Fieseler, L., Klumpp, J. and Loessner, M. J. (2012). Biocontrol of *Salmonella* Typhimurium in RTE foods with the virulent bacteriophage FO1-E2. *International Journal of Food Microbiology*, (article in press).

Hagens, S. and Loessner, M. J. (2007). Application of bacteriophages for detection and control of foodborne pathogens. *Applied Microbiology and Biotechnology,* 76, 513-519.

Holck, A. and Berg, J. (2009). Inhibition of *Listeria monocytogenes* in cooked ham by virulent bacteriophages and protective cultures. *Applied and Environmental Microbiology*, 75 (21), 6944–6946.

Katsuda, K., Hata, E., Kobayashi, H, Kohmoto, M., Kawashima, K., Tsunemitsu, H. and Eguchi, M. (2005). Molecular typing of *Staphylococcus aureus* isolated from bovine mastitic milk on the basis of toxin genes and coagulase gene polymorphisms. *Veterinary Microbiology,* 105, 301-305.

Kennedy, J. E. J. and Bitton G. (1987). Bacteriophages in foods. In: Goyal, S. M., Gerba, C. P., Bitton, G. (Eds.), Phage Ecology. New York: John Wiley and Sons, pp. 289–316.

Kim, K., Klumpp, J. and Loessner, M. (2007). *Enterobacter sakasakii* bacteriophages con prevent bacterial growth in reconstituted infant formula. International *Journal of Food Microbiology*, 115, 195-203.

Kobayashi, H., Shimada, J., Nakazawa, M., Morozumi T., Pohjanvirta T., Pelkonen, S. and Yamamoto, K. (2001). Prevalence and characteristics of Shiga toxin-producing *Escherichia coli* from healthy cattle in Japan. *Applied and Environmental Microbiology*, 67,484–489.

Kudva, I., Jelacic, S., Tarr, P. I., Youderian, P. and Hovde, C. J. (1999). Biocontrol of *Escherichia coli* O157 with O157-Specific Bacteriophages. *Applied And Environmental Microbiology*, 65 (9), 3767–3773.

Le Loir, Y., Baron, F. and Gautier, M. (2003). *Staphylococcus aureus* and food poisoning. *Genetics and Molecular Research*, 2, 63-76.

Mead, P. S., Slutsker, L., Dietz, V., McCaig, L. F., Bresee, J. S., Shapiro, C., Griffin, P. M. and Tauxe, R. V. (1999). Food-related illness and death in the United States. *Emerging Infectious Diseases*, 5, 607–625.

Modi, R, Hirvi, Y., Hill, A. and Griffiths, M. W. (2001). Effect of pahge on survival of *Salmonella enteritidis* during manufacture and storage of cheddar cheese made from raw and pasteurized milk. *Journal of Food Protection,* 64, 927-933.

Moradpour, Z., Sepehrizadeh, Z., Rahbarizadeh, F., Ghasemian, A., Yazdi, M. T. and Shahverdi, A. R. (2009). Geneticallyengineered phage harbouring the lethal catabolite gene activator protein gene with an inducer-independent promoter for biocontrol of *Escherichia coli. FEMS Microbiology Letters,* 296, 67–71.

Obeso, J. M., Garcia, P., Martínez, B., Arroyo-López, F. N., Garrido-Fernández, A. and Rodriguez, A. (2010). Use of logistic regression for prediction of the fate of *Staphylococcus aureus* in pasteurized milk in the presence of two lytic phages. *Applied and Environmental Microbiology,* 76 (18), 6038-6046.

O'Flynn, G., Ross, R. P., Fitzgerald, G. F. and Coffey, A. (2004). Evaluation of a cocktail of three bacteriophages for biocontrol of *Escherichia coli* O157:H7. *Applied and Environmental Microbiology*, 70, 3417–3424.

Pearson, L.J. and Marth, E.H. (1990). *Listeria monocytogenes*--threat to a safe food supply: a review. *Journal of Dairy Science*, 73 (4), 912-928.

Petty, N. K., Evans, T. J., Fineran, P. C. and Salmond, G. P. (2007). Biotechnological exploitation of bacteriophage research. *Trends in Biotechnology,* 25 (1), 7-15.

Ritchie, Jenny M., Wagner, Patrick L., Acheson, David W. K. and Waldor, Matthew K. 2003. Comparison of Shiga toxin production by hemolytic-uremic syndrome-associated and bovine-associated Shiga toxin-producing *Escherichia coli* isolates. *Applied and Environmental Microbiology*, 69, 1059–1066.

Rohwer, F. and Edwards, R. (2002). The Phage Proteomic Tree: a Genome-Based Taxonomy for Phage. *Journal of Bacteriology,* 184 (16), 4529–4535.

Schultze W. D. and Brasso W. B. (1987). Characterization and identification of *Mycobacterium smegmatis* in bovine mastitis. *American Journal of Veterinary Research*, 48(5), 739-742.

Su, C. and Brandt, L. J. (1995). *Escherichia coli* O157:H7 infection in humans. *Annals of Internal Medicine*, 123, 698–714.

Sulakvelidze, A. and Barrow, P. (2005). Phage therapy in animals and agribusiness. In: E. Kutter, A. Sulakvelidze, p. 335–380. (Eds.), Bacteriophages: Biology and Applications. Boca Raton: CRC Press.

Tasara, T. and Stephan, R. (2006). Cold stress tolerance of *Listeria monocytogenes*: A review of molecular adaptive mechanisms and food safety implications. *Journal of Food Protection*, 69 (6), 1473-1484.

Thorpe, C. M., Ritchie, J. M. and Acheson, D. W. K. (2002). Enterohemorrhagic and other Shiga toxin-producing *Escherichia coli.* In M. Donnenberg (ed.), *Escherichia coli*: virulence mechanisms of a versatile pathogen. Academic Press, San Diego, California.

Tirado, C. and Schmidt, K. (2000). WHO Surveillance Programme for Control of Foodborne Infections and Intoxications in Europe. Seventh Report. (http://www.bfr.bund.de/internet/7threport/7threp_fr.htm).

Vashin, I. and Stoyanchev, T. (2005). Presence of *Campylobacter* spp. in meat and internal organs of japanese quail (Coturnix Coturnix). *Trakia Journal of Sciences,* 3 (5), 23-25.

Waldvogel, F.A. (2001). *Staphylococcus aureus* (including Staphylococcal toxic shock), In: Mandell, J.L., Bennet, J.E. and Dolin, R. (Eds.), Principles and practice of infectious diseases. Churchill Livingstone, Philadelphia, pp. 2069-2092.

WHO. (1998). Zoonotic non-O157 Shiga toxin-producing *Escherichia coli* (STEC). Report of a WHO Scientific Working Group Meeting. World Health Organization, Geneva, Switzerland.

Wiggins, B. A. and Alexander, M. (1985). Minimum bacterial density for bacteriophage replication: implications for significance of bacteriophages in natural ecosystems. *Applied and Environmental Microbiology*, 49, 19-23.

Zuber, S., Boissin-Delaporte, C., Michot, L., Iversen, C., Diep, B., Brüssow, H. and Breeuwer, P. (2008). Decreasing *Enterobacter sakazakii* (*Cronobacter* spp.) food contamination level with bacteriophages: prospects and problems. *Microbial Biotechnology*, 1(6), 532-543.

In: Bacteriophages
Editors: Clark Denton and Richard J. Crosby
ISBN: 978-1-62618-513-5

Chapter 5

Bacteriophages As Probiotics and Decontaminating Agents for Food Products

Andrey V. Aleshkin,[1,*] Nikolay V. Volozhantsev,[2,†] Edward A. Svetoch[2] and Stanislav S. Afanas'ev[1]
[1]Gabrichevsky Moscow Research Institute of Epidemiology and Microbiology, Russian Federation
[2]State Research Center for Applied Microbiology and Biotechnology, Obolensk, Russian Federation

Abstract

Bacteriophages are components, determinable quantitatively, of natural microflora of food production continuum which covers all production stages: from growing agricultural plants and animals to retail sales of ready food products. The viruses of bacteria are stably present in all objects of the environment and may be easily extracted from soil specimens, fresh and sea water, washouts from plants surfaces, human and animals excrements, ready for sale food products. Sterile and cleansed of bacterial toxins the phage lysates with high titers of viral particles are used for species-specific decontamination of food products in all stages of manufacturing process and storage and are employed as probiotic dietary supplement that allow to reduce the risk of food-borne infections caused by *Escherichia*, *Salmonella*, *Shigella* and other pathogenic bacteria, including their antibiotic resistant strains. Protection of food products from bacterial contamination using the strategy of phage biocontrol is a highly differentiated preserving technique which has prolonged effect, is safe for humans and is relatively inexpensive as compared to commonly used techniques.

* Corresponding author: Andrey V. Aleshkin, Ph.D., Sci.D. (Biol.), Laboratory of Bacteriophage Biotechnology, Gabrichevsky Moscow Research Institute of Epidemiology and Microbiology, Russian Federation, E-mail: ava @gabri.ru.

† Corresponding author: Nikolay V. Volozhantsev, Ph.D., Department of Microbiology, State Research Center for Applied Microbiology and Biotechnology, Obolensk, Russian Federation, E-mail: nikvol@obolensk.org.

Keywords: Bacteriophage, phage therapy, bacteriophage biocontrol, bacteriophage bioprocessing, phage-based probiotic dietary supplement

Bacteriophages are viruses which posses a specific capability to infect selectively bacteria that belong to a certain strain or antigenic homologous strains of a certain species or genus, and subsequently, after intracellular replication to lyse host cells (virulent phages) or become integrated into a bacterial genome forming lysogenes (temperate phages) [1]. According to the data reported phages can be used as natural antimicrobial agents to control bacterial infections in humans, animals and agricultural plants [2-4]. The whole range of such medical treatments is labeled by specialists as phage therapy [5]. A number of authors consider that it is feasible to use bacteriophages in conjunction with sanitary and hygienic activities in food industry, public catering, children's collectives, military units and hospitals [6-11]. All such possible applications of bacteriophages were merged by specialists into one category termed phage-mediated biocontrol, by this term specialists mean that phages can be used directly by humans in the peroral form (phage-based probiotic dietary supplement) or in some other forms of phage compositions (tooth paste, cream, deodorant, etc.), as well as to use phages to control bacterial pathogens in agricultural plants and animals (before harvest and slaughter), to treat instruments and equipment in hospitals and food industry plants and to process half-cooked food or ready-to-eat (RTE) products in order to reduce the number of certain pathogenic bacterial strains present in these facilities, including those that cause food-borne infections [12, 13]. In this survey, first of all, we shall consider the issues of safe and effective application of bacteriophages as processing aid (decontaminating agents of food products in all production stages) and probiotic dietary supplement that are an additional valuable source of bacteriophages for human organism; their presence in human normoflora allows to reduce the risk of such widely spread food-borne infections in humans as salmonellosis, campylobacteriosis, dysentery, escherichiosis, etc.

Bacterial viruses were discovered by Frederick Twort in 1915, but the era of bacteriophages began after papers on the subject were published by Felix d'Herelle, who was the first to demonstrate the presence of phages in normal microflora of humans and animals 95 years ago [1, 14]. Many subsequent investigations confirmed the presence and isolation of phages from feces, urine, saliva and sputum of healthy and ill humans and animals when their concentration reached 10^6 pfu/ml [15-23]. In addition, it was shown that the bacterial viruses are present everywhere in the objects of the environment, water, soil, plants, etc. in amounts exceeding several million particles per unit of substrate (for example, in a drop of sea water) [24]. Thus, the issue of bacteriophages application in human diet may be considered more likely in quantitative terms (taking into account viruses count in the product), than in qualitative terms, as it is examined in the case of chemical antibacterial remedies. Some authors placed bacteriophages into the class of probiotic, pointing out that their biological nature is completely consistent with the WHO definition which reads: "Probiotics are live microorganisms which, when administered in adequate amounts, confer a health benefit on the host" [25-27]. The specificity of phage particle interaction with the indicator bacterial culture in principle limits the possibility of bacteriophage direct negative action on human organism (cells) that may be used for safe decontamination of products and prevention of bacterial carriage. But it may be suggested theoretically that the administration of bacteriophages can produce some side-effects in humans:

1 Infection of bacteria of normal human flora accompanied by iatrogenic dysbacteriosis, as is the case in the administration of other antimicrobial agents. The selection of bacteriophage production strains which are used for phage therapy allows to isolate viruses particles with the narrowest possible range of lytic activity limited within the species or even the strain of bacterial host [28, 29]. Such selection allows to exclude the penetration of viruses that may damage bifidobacteria and lactobacilli when designing phage cocktails which are used to reduce the risk of food-borne infections [24]. The administration of mono- and polyvalent phage preparation for the treatment of acute intestinal infections and decompensated forms of dysbacteriosis in Russia (Soviet Union) and Poland over more than 50 years also confirmed that mentioned preparations cannot lyse the normal human microflora [30-33].

2 Stimulation of immune reaction in humans. The results of conducted clinical researches showed that bacteriophages exert influence on different functions of the main populations of cells of human immune system which participate in forming both the innate and adaptive immunity: production of cytokines, proliferation of T-cells, synthesis of antibodies and finally phagocytosis and respiratory bursts of phagocytes [33, 34]. At the same time the local and systemic phage therapy administered to patients produced no significant anaphylactic reactions, excluding the endotoxic one that was owed both to the quality of used phage composition (it will be examined later) and to the reaction of bacteriolysis (Jarisch-Herxheimer reaction) in situ [35]. Minimizing the concentration of phage particles in prophylactic bacteriophagic treatment to the titer equal to the isolated from the human organism (10^3-10^6 pfu/ml), it is possible to decrease the amount of endotoxin that is produced instantaneously in the bacteriolysis [35]. The Polish researchers have called attention to the modulating influence of bacteriophages on some immune reactions which are caused by pathogenic bacteria and viruses that may be manifested positively in the synergetic (antibacterial and immunomodulating) effect of bacteriophages [33, 34].

3 Modulation of virulence of bacteria host owing to the modification of the bacterial phenotype because of the penetration of prophage genome into the host DNA (the genome contains genes of virulence) or the transduction of such genes (for example, toxin-encoding genes) from lysogenic pathogenic bacterial cultures to non-pathogenic ones (so called horizontal transfer of genes). To avoid such undesirable effects of bacteriophages on a human organism, directional selection of production strains should be applied, implying incorporation of such lytic phages into the composition which will never form stable lysogenes of a bacterial culture [36-38]. Moreover, using present-day PCR techniques and DNA sequencing it is possible to control the presence of toxin-encoding genes and other factors of virulence both in the bacteriophage genome and in the bacterial host genome, and, hence, to ensure the most safe administration of bacteriophages for humans [12, 22].

4 Introduction, into human organism, of bacteria-host toxins which are present in the sterile filtrate of phage lysate. The quality of ready phage cocktail is tested for contents of endo- and exotoxins [39]. The concentration of exotoxins may be reduced to zero using non-pathogenic bacterial cultures in order to increase the biomass of bacterophages (for example, *E. coli* K-12 for Escherichia and *L. innocua* for Listeria phages) [40]. The content of endotoxin in the composition is determined after the

additional purification of the sterile filtrate of phage lysate and is expressed in units of endotoxin per 1 ml according to the regulating rules of the corresponding producing country (thus, according to the documents of the European Food Safety Authority (EFSA) it is 50 EU/ml) [36]. The quantity of the endotoxin isolated in situ is not tested, but bacteriolysis in the intestine may be regulated using the dosage and number of administrations of phage cocktail.

Thus, investigating phages and their interactions with eukaryotic cells (including animal and human) over more than 90 years, researchers have not found evidence that bacteriophages can produce negative specific effects on human health [5].

Considering the efficiency of bacteriophage application in humans it should be mentioned that more than 40 years ago in the former Soviet Union, first the G. Eliava Tbilisi Research Institute for Vaccines and Sera and I.I. Mechnikoff Ufa Research Institute for Vaccines and Sera and later in the Russian Federation the subsidiaries of the company Microgen (in the cities of Nizny Novgorod, Perm and Ufa) began to produce more than ten brands of medicinal agents based on one species of bacteriophages or their combinations. These preparations were used to treat and prevent acute intestinal infections and decompensated forms of dysbacteriosis and to control causative agents of some suppurative infections [30, 41]. They are liquid and tablet forms of medicinal monovalent preparations with *Staphylococcus*, *Streptococcus*, *E. coli*, *Klebsiella*, *Salmonella*, dysentery, typhoid, *Proteus* and pyocyaneum bacteriophages or cocktails which contain several kinds of phages: Coli-proteus®, Pyobacteriophage combined® (against *Staphylococcus* spp., *Streptococcus* spp., *Proteus* spp. and *E. coli*), Polyvalent pyobacteriophage® or Sexstaphage® (against *Staphylococcus*, *Streptococcus*, *Klebsiella*, *Proteus*, *P. aeruginosa* and *E. coli*), Intestiphage® (against *Shigella* spp., *Salmonella* spp., *Staphylococcus* spp., *Enterococcus* spp., *Proteus* spp., *P. aeruginosa* and *E. coli*) [42]. The medicinal preparations are sterile filtrates of bacterial phage lysates, they are intended for internal and local use: irrigation of wounds and mucosas, introduction into cavity of uterus, bladder, ear, paranasal sinus, bulbar conjunctiva as well as into drained cavities (abdominal and pleural), into abscess cavities after the pus is removed (Table 1).

In recent years in Russia bacteriophages have found appropriate use as therapeutic agents in dental and dermatological practice. Toothpastes and creams based on bacteriophages are produced commercially [43].

Global therapeutical practices involving phages seem to be far less popular than those in Russia (probably except Poland and Georgia). While tens of thousands of patients are given bacteriophages on doctor's orders annually in Russia (which correlates quantitatively with the volume of sales of medicinal preparations based on bacteriophages, according to the data of annual retail audit performed by the companies IMS Health, DSM, RMBC and Pharmexpert) within the framework of the Russian Ministry of Health standard: «Intestinal dysbacteriosis. Protocol for the treatment of patients» [44], the few today's reports of Western colleagues about initial stages of tests on humans, conducted in the US, Great Britain and Belgium, do not look impressive (Table 2). It is a well-known fact, however, that before the introduction of penicillin in 1940, bacteriophages were produced on a mass scale by L'Oreal in France (5 brands), and by Eli Lilly in the US (6 brands of mono- and polyvalent phage-based medicinal agents) [14]. On the other hand, the information on expertise in using phages in conjunction with the sanitary epidemiological procedures in the EU member states, US and Canada is extensively published in the scientific literature [9, 12, 13].

Such measures include four areas of activities: (1) bacteriophage-mediated biocontrol that means the use of bacteriophages to control bacteria that are damaging agricultural plants and animals during the stages that precede their processing in food industry (fruit and vegetables before harvesting, and cattle and poultry before slaughtering); (2) phage bioprocessing means the use of bacteriophages to decontaminate vegetables, fruit, meat, fish, etc. during their industrial processing before packaging finished products. (3) prophylactic administration of bacteriophages by humans as probiotic dietary supplement to reduce the risk of sporadic cases and epidemic outbreaks of food-borne infections. (4) phage-based detection and identification of potentially dangerous microorganisms (not reviewed in this Chapter).

An example of the phage-mediated biocontrol is the biopesticide AgriPhage® which contains a cocktail of virulent phages that are lysing *Xanthomonas campestris* pv. *vesicatoria* and *Pseudomonas syringae* pv. *tomato*, which damage tomato and pepper fruits (so called "bacterial black spot"). It was registered by the US Environment Protection Agency (EPA) in December 2005 [37]. Originally the bacteriophages comprising the biopesticide were isolated from vegetables infected with these bacteria. One liter of phage cocktail produced by the US company OmniLytics, contains not less than 4.1×10^{12} phage particles (that is 4.1×10^{9} pfu/ml). To spray 1 acre of vegetable plantations a farmer has to use one to two pints of AgriPhage® diluted in 100 gallons of water [60].

In 2006-2007 OmniLytics got authorization from the US Department of Agriculture (DA) to use compositions based on bacteriophages for the control of *E. coli* O157:H7 and *Salmonella* (BacWash®) that is necessary for treating skins of domestic cattle and all places where animals may be infected before slaughtering (enclosures, pens, transportation containers, etc.) [33, 60].

The Table 3 shows the data of investigations that evidenced high effectiveness of bacteriophages for biocontrol of plants and animals which are used later as food for humans.

As an example of phage bioprocessing we can cite a phage-based processing aid registered by US Food and Drug Administration FDA in 2006 which consists of polyvalent listeriosis cocktail ListShield® (LMP-102, produced by Intralytix, US) – «Listeria monocytogenes Specific Phage Preparation», that contains six different phages. 170 strains of L. monocytogenes are sensitive to this processing aid that allows to treat RTE products made of meat of farm animals and poultry.

This processing aid is used immediately before packing in meat processing factories and poultry farms in order to decontaminate half-cooked foods, sausages, meat cuts using aerosol to spray their surfaces at a rate of not more than 1 ml per 500 cm^2 of food product surface [10, 38]. The efficiency of this cocktail in the processing of fruits was demonstrated experimentally [61, 62].

In the same year the FDA approved processing aid Listex® (produced by Micreos, Netherlands) based on Listeria bacteriophage strain P100 as Generally Recognised as Safe (GRAS) product.

First it was used for producing cheese, and later in 2007 for other food products. EFSA granted authorization to decontaminate fresh fish with Listex® in 2012 [12, 22, 36, 76]. The Listeria bacteriophage is used in bioprocessing of different RTE foods such as soft cheese, hot dog, smoked salmon, mixed seafood, cabbage and others [40, 79].

Table 1. Data on human bacteriophage therapy studies in Russia and the former Soviet Union

Author and publication year	Name of bacteriophage preparation	Etiologic agent	Nosology	Number of patients	Efficiency of bacteriophage therapy	Details of research	Ref.
Akimkin et al., 2010	Salmonella bacteriophage®	*Salmonella*	Nosocomial salmonellosis	459 patients and 2500 hospital staff	No facts of bacterial shedding in the hospital for 12 months after phage prophilaxis	2-4 tablets given 3 times a day for 5-7 days. The facts of self-recovery were reported because patients had their own bacteriophage strains in their organisms	[6]
Anpilov et al., 1984	Polyvalent dysentery bacteriophage®, in tablets with acid-proof coating	*Shigella flexneri*	Dysentery	3000 persons	Effective remedy to prevent outbreaks in enclosed collectives	2 tablets given once every 3 days from June to October in different geographic zones of the Soviet Union	[7]
Babalova et al., 1968	Dysentery bacteriophage®, in tablets	*Shigella flexneri and S. sonne, E. coli*	Dysentery	30769 children at the age of 0.5 to 7 years	Number of diseased persons in study group is 3.8 times lower than in the control group	1 tablet given every week for 109 days	[45]
Voroshilova et al., 2010	Polyvalent Enterobacter purified®	*E. aerogenes, E. cloacae, E. agglomerans*	Intestinal infections	110 children	More effective than traditional antibiotic therapy by 20-40%		[46]
			Septic diseases of neonates	26 children			
			Urological infection	50			
			Surgical infection	20			
Zhukov-Verezhnikov et al., 1978	Adapted individually production strains of *Staphylococcus, Streptococcus, E.coli, Proteus, P. aerugenosa phages*	*Staphylococcus, Streptococcus, E.coli, Proteus, P.aerugenosa*	Surgical purulent infection	60	Use of bacteriophage preparations was highly effective	Adapted individually strains are more efficient than production ones.	[47]

Author and publication year	Name of bacteriophage preparation	*Etiologic agent*	Nosology	Number of patients	Efficiency of bacteriophage therapy	Details of research	Ref.
Zaharova et al., 2010	Coliphages®, Staphylophages®, Streptococcus bacteriophage®	*E. coli, S. pyogenes, S. aureus, K. pneumoniae*	Urinary tract infection (UTI)	65 pregnant women	Administration of phages to pregnant women with UTI allowed to prevent complications in pregnancy and child-birth, acute purulent septic diseases in mothers and neonates and to reduce maternal and child mortality	Oral dose of 20.0 ml of preparation 2-3 times a day before breakfast and simultaneously 5-10.0 ml of the preparation administered in pads intravaginally for 40-60 minuites. The course lasts 7-10 days	[42]
Litvinova et al., 1978	Coli-proteus bacteriophage®	*E. coli, Proteus*	Neonatal sepsis	500 premature children	Administration of bacteriophage allowed to reduce the lethality	Dose of 2 ml/kg administered twice a day during 3 days. The course is repeated after the interval of 3-4days	[48]
Milutina, 1990	Salmonella bacteriophage®, Dysentery bacteriophages®	*S.* Typhimurium, *S.* Enteritidis, *Shigella flexneri*	Salmonellosis, dysentery	45 children under the age of 12 months; 67 children above the age of 12 months	Recovery effect of salmonellosis phage is 81.8-85.4 %, and that of dysentery phage, 90.9%	Synergism of phage and antibiotic therapy was reported	[49]
Nilova et al., 2007	Staphyphage®, Proteus bacteriophage®, Coliphage®, Aeruginosaphage®, Coli-proteus®, Pyopolyphage®, Sextaphage®, Intesti bacteriophage®, Klebsiella bacteriophage®, Klebsiella pneumonia polyvalent®	*S. aureus, E. coli, Klebsiella spp., Proteus spp., P. aeruginosa, E. faecalis*	Dysbacteriosis	80 children under the age of 12 months, 47 children at the age of 1 to 3 years	Phage therapy is promising method to correct the dysbacteriosis of large intestine	Tests were performed to determine the sensitivity of conditionally pathogenic microorganisms to preparations of different producers (subsidiaries of Microgen)	[50]

Table 1. (Continued)

Author and publication year	Name of bacteriophage preparation	Etiologic agent	Nosology	Number of patients	Efficiency of bacteriophage therapy	Details of research	Ref.
Pogorelskaya et al., 1968	Coli-proteus bacteriophage®, Dysentery bacteriophage®	Enteropathogenic intestinal bacilli: *Pr. vulgaris, Pr. mirabilis, Shigella flexneri*	Acute intestinal infection	3547 children at the age of 1 to 3 years	Phage prophylaxis during epidemic season reduced the incidence of intestinal infections.	Multiple regular administration of phages (16-20 times a year)	[28]
Shahmardanov et al., 1997	Pyobacteriophage polyvalent®	*Shigella flexneri*	Dysentery,	29	Reduction in the length of diarrhea and alleviation of pains in the abdomen as compared with the standard therapy	Daily dose is 30 ml	[51]
			salmonellosis	41			
Solodovnikov et al., 1970	Polyvalent dysentery bacteriophage®, in tablets with acid-proof coating	*Shigella flexneri and S. sonne*	Dysentery, acute intestinal infection	6522 children at the age of 1 to 7 years	Number of diseased persons in study group is 2.1-2.5 times lower than in the control group	1-2 tablets given every week from June to October	[52]

Table 2. Data on human phage therapy studies in the world (excepting Russia)

Authors and publication year	Name of the preparation	Nosology	Number of patients	Efficiency of phage therapy	Details of research	Ref.
Markoishvili et al., 1999	PhagoBioDerm®	Skin sores owing to venous congestion	100	Full cicatrisation of sores in 70% of cases	Combined preparation of prolonged effect based on polymer impregnated with bacteriophage and antibiotic	[53]
Slopek et al., 1987	Phage strains selected individually	Varicosity with ulceration and inflammation	36	Positive result of phage therapy in 75% of cases		[54]
		Furunculosis and inflammation of connective tissue and lymphatic vessels	162	Recovery in 95%		
		Purulent infection	370	Positive effect in 92.4% of cases		
		Purulent pericarditis	7	100% of efficiency		

Authors and publication year	Name of the preparation	Nosology	Number of patients	Efficiency of phage therapy	Details of research	Ref.
Slopek et al., 1987	Phage strains selected individually	Methicillin resistant staphylococcus infection	550 adults, 90 children	93% of efficiency 95.5% of efficiency		
		Pyogenic arthritis and myositis	19	Positive effect in 89.5% of cases		
		Pyogenic osteomyelitis	40	90% of efficiency		
		Diseases of gastrointestinal tract (including food-borne infections)	48	Efficiency of phage therapy was 100%		
		Diseases of respiratory system	180	Positive dynamics was in 86.7% of cases		
		Conjunctivitis blepharoconjunctivitis, otitis media	16	Efficiency of phage therapy was in 93.8%		
		Diseases of urogenital tract	42	Positive dynamics was in 92.9% of cases		
		Septicemia	98	Recovery was reported in 88.8% of cases		
Weber-Dabrowska et al., 2003	Bacteriophage strains selected individually	Septic infection	94	80 patients had full recovery, 14 patients had only lower temperature (unsuccessful treatment)		[55]
Cislo et al., 1987	Bacteriophage strains selected individually	Chronic pustular infection of the skin	31	25 patients made improvement owing to phage therapy	Oral and local administration of phages was used	[56]
Abdul-Hassan et al., 1990	Phage strains against P. aeruginosa previously were isolated from the wound	Wound burn infection that produced antibiotic resistant sepsis (*P. aeruginosa*)	30	24 patients made improvement	Treatment consists in changing 3 times a day a bandage impregnated with phages which titer is 10^{10} pfu/ml	[57]
Merabishvili et al., 2009	BFC-1®	Wound burn infection caused by *S. aureus, P. aeruginosa*	8	No side effects were reported	Phage was applied to the wound surface in the form of aerosol	[29]
Lang et al., 1979	Bacteriophage strains selected individually	Septic complications in orthopedic surgery	7	Efficiency was reported in 6 cases		[58]
Wright et al. 2009	Biophage-PA®	Chronic otitis caused by *P. aeruginosa*	24	Full recovery in 25% of cases, improvement in 92% of cases		[59]

Table 3. Examples of bacteriophage-mediated biocontrol in plants and animals which are used to produce food products ([9] with some additions)

Species of plant or animal	Controlled infection	Bacteriophage host strain	Ref.
Cultivated mushrooms	Bacterial blotch	*Pseudomonas tolaasii*	[63]
Tomatoes	Bacterial spot	*Xanthomonas campestris pv. vesicatoria*	[64]
Apples	Fire blight	*Erwinia amzlovara*	[65]
Stone fruits	Prunus bacterial spot	*X. campestris pv. pruni*	[66]
Sprouts	Contamination of seeds	*Salmonella* Enteritidis	[67]
Fish	Red fin disease	*Aeromonas hydrophila*	[68]
	Hemorrhagic ascites	*Pseudomonas piecoglossicida*	[69]
Chicken	Cecal Salmonella	*Salmonella* Enteritidis	[70]
	Lethal infection	*Salmonella* Typhimurium	[71]
	Respiratory infection	*Eschirichia coli*	[72]
Beef cattle	Bacterial shedding	*E. coli* O157:H7	[73]
Diary cattle	Mastitis	*Staphilococcus aureus*	[74]
Pigs	Tonsil and cecal Salmonella	*Salmonella* Typhimurium	[75]

It is significant that the effective concentration of the bacteriophage depends on the decontaminated product.

For example, the phage particles are spreading uniformly and freely in liquid food products (milk and cheese brine). A different situation arises with food products that have even solid surface (hot dogs, lettuce leaves) because low capability to absorb the bacteriophage suspension makes it difficult to decontaminate the product. From the phage bioprocessing standpoint, the most complicated products are those with uneven and large surface (fish, meat and seafood) that limits physically the transportation of phage particles to the bacterial cells-targets.

Thus, striving to maximize the effect of decontamination on food products it has been found experimentally that the concentration of the Listeria bacteriophage suspension which may be used for all kinds of products should not be less than 10^8 pfu/ml or cm^2 [77].

In 2011 Intralytics submitted for the authorization of the US FDA a cocktail of 3 virulent bacteriophages (EcoShield® or ECP-100) belonging to the family of Myoviridae, which originally were isolated from sewage and sea waters. They are active against 111 strains of *E. coli* O157:H7, 76 strains of other *E. coli* serotypes and 20 strains of other bacterial species. The company was granted the Food Contact Notification that allows to use the cocktail in processing of raw meat and ground beef [78, 12]. The authors demonstrated high efficiency of processing aid (number of bacteria E. coli decreased to 5% during 24 hours) developed for the decontamination of the allowable group of half-cooked products [78].

In the same year the Netherlands Medical Expert Council granted to Micreos a provisional authorization to use a processing aid based on the Salmonella bacteriophage (Salmonelex®) in order to conduct full-scale trials in one of the largest poultry processing factories in EU [76]. In 2012 in order to obtain authorization for the GRAS product

SalmoFresh® the company Intralytics presented for the consideration of the FDA a cocktail of lytic bacteriophages specifically active against highly pathogenic *Salmonella* strains of the following serotypes: Typhimurium, Enteritidis, Heidelberg, Newport, Hadar, Kentucky and Thompson. The cocktail was developed specifically for decontamination of raw meat and poultry before butchery [79]. The experimental research of independent investigators showed that using the artificial infection of the chicken skin with *S.* Enteritidis and chicken legs and sausages of chicken meat with *S.* Typhimurium, the phage cocktail demonstrated effectiveness and allowed to reduce the bacterial population by 2 lg [12, 13, 9]. Using bacteriophage cocktail (PC1) containing the phage Felix 01, active against *S.* Typhimurium, with the concentration more than 10 times exceeding *Salmonella* level it was possible to reduce bacterial population of pig's skin by 99% [80].

The third sphere of sanitary epidemiologic use of bacteriophages includes activities that allow to reduce significantly the risk of outbreaks and sporadic cases of food-borne infections. These requirements are completely met by the probiotic dietary supplement based on Shigella phage (commercial name ShigActive®) which was developed by the company Intralytics. Presently the third stage of the project is being implemented with the aim to use extensively this dietary supplement in the US army [79]. The idea of administering *Shigella* and other species of bacteriophages (*Salmonella* etc.) for prophylaxis of food-borne infections is not a new one and was used in the Soviet Armed Forces and other collectives (schools, hospitals etc.) during the Soviet era [7, 52, 28]. But until recently the bacteriophage preparations were used in the form of medicine. The idea of introducing bacteriophages in the diet and using them as probiotic dietary supplements was proposed by American scientists [25-27].

In conclusion it is pertinent to note that presently there are a dozen companies around the world which are producing medicine, veterinary remedies, processing aids and probiotic dietary supplements based on bacteriophages (Table 4).

The increase in antibiotic resistant strains of *Staphylococcus*, *P. aeruginosa, Campylobacter, Clostridia, Acinetobacter* and many others is an unabated incentive for these companies to create new phage preparations active against pathogenic microorganisms mentioned above.

Table 4. Producers and developers of phage-based preparations

Name of the company	Key competence	Country and Web site address
Biochimpharm or Biopharm Pharmaceuticals	Production in the G. Eliava Institute (mono- and polyvalent phages: staphylococcus, streptococcus, coliphage, intestiphage, pyophage, etc.)	Georgia www.biochimpharm.ge
AmpliPhi Biosciences (before Biocontrol and Targeted Genetics)	Clinical trials of pyocyaneum phage BioPhage-PA® (treatment of otitis) and BioPhage-RA® (cystic fibrosis)	US and UK www.ampliphibio.com
Biophage Pharma Inc.	Clinical trials on pigs and birds with *E. coli* and *S.* Typhimurium phages [81]	Canada www.biophagepharma.net
GangaGen Biotechnologies	Development of phage-based preparations against MRSA strains *S. aureus* and *P. aeruginosa*	US, India www.gangagen.com

Table 4. (Continued)

Name of the company	Key competence	Country and Web site address
Microgen	Large scale production in 3 plants (mono- and polyvalent phages: staphylococcus, streptococcus, pyocyaneum, proteus, coliphage, intestiphage®, pyocombined®, sextaphage®, etc.)	Russia www.microgen.ru
Intralytix	Probiotic dietary supplement based on *Shigella* phage ShigActive®, polyvalent *Listeria* phage cocktail ListShield® to process ready-to-eat products, polyvalent *E. coli* phage cocktail to process raw meat and ground beef EcoShield®, cocktail of lytic *Salmonella* bacteriophages to process raw meat and poultry SalmoFresh®	US www.intralytix.com
Micreos	Listex®, polyvalent *Listeria* phage P100 to process any kind of half-cooked and RTE products. Salmonelex®, *Salmonella* phage to process raw meat and poultry	Neththerlands www.micreos.com
OmniLytics	Biopesticide AgriPhage®, containing cocktail of virulent phages that lyse *Xanthomonas campestris pv. vesicatoria* and *Pseudomonas syringae pv. tomato*, which damage tomato and pepper fruits (so called black bacterial spot)	US www.omnilytics.com
Novolytics	Development of phages against MRSA strains *S. aureus* for local use and for treatment of catheters and other devices that are introduced into organism, based on gel	UK www.novolytics.co.uk
Phage Biotech	Clinical trials of eye and ear drops based on *P. aeruginosa* phage	Israel www.phage-biotech.com
Special Phage Services	Clinical trials of MediPhage®, staphylococcus bacteriophage (MRSA) strains and development of phage cocktails against polyresistant *P. aeruginosa* (VRPA) and vancomycin resistant enterococcus (VRE)	Australia www.specialphageservices.co m.au
Viridax	Development of phages against *S. aureus* and other *Staphylococcus*	US www.viridax.com

References

[1] Adams, H. M. (1959). *Bacteriophages*. New York: Interscience Publishers Inc., London: Interscience Publishers ltd.

[2] Krasil'nikov, I. V., Lobastova, A. K. and Lysko, K. A. (2010). Some aspects of the current state and future directions of bacteriophage preparations' production and application. *Biopreparaty*, 2(38), 28-33 [in Russian].

[3] Alisky, J., Iczkowski, K., Rapoport, A., and Troitsky, N. (1998). Bacteriophages show promise as antimicrobial agents. *J. Infect.*, 36, 5-15.

[4] Brussow, H. (2005). Phage therapy: the Escherichia coli experience. *Microbiology*,151, 2133-2140.

[5] Letarov, A., Golomidova, A. K. and Tarasyan, K. K. (2010). Ecological basis of rational phage therapy. *Acta Naturae*, 2, 60-71.

[6] Akimkin, V. G., Darbeyeva, O. S. and Kolkov, V. F. (2010). Historical and modern aspects of bacteriophages application: Experience and prospects. *Klinicheskaya praktika*, 4, 48-54 [in Russian].

[7] Anpilov, L. I. and Prokudin, A. A. (1984). Prophylactic efficiency of dry polyvalent dysentery bacteriophage in structured collectives. *Voyenno-meditsinskiy zhurnal,* 5, 39-40 [in Russian].

[8] Drozdova, O. M. and Brusina, Ye. B. (2010). Use of bacteriophages in epidemiological practice: a view after a century. *Epidemiology and Infection diseases*, 5, 20-24 [in Russian].

[9] Greer, G. G. (2005). Bacteriophage control of foodborne bacteria. *J. Food Protection,* 68 (5), 1102-1111.

[10] Peek, R. and Reddy, K. R. (2006). FDA approves use of bacteriophages to be added to meat and poultry products. *Gastroenterology*, 131, 1370-1372.

[11] Walker, Kathy. (2006). Use of bacteriophages as novel food additives. FS06 ANR 490/811 *Food Regulation in the United States*, Michigan State University.

[12] Goodridge, D. L. and Bisha, B. (2011). Phage-based biocontrol strategies to reduce foodborne pathogens in foods. *Bacteriophage*, 1 (3), 130-137.

[13] Goodridge, D. L. and Abedon, T. S. (2003). Bacteriophage biocontrol and bioproccessing: application of phage therapy to industry. *SIM News*, 53 (6), 254-262.

[14] Stone, R. (2002). Stalin's forgotten cure. *Science*, 298, 728-731.

[15] Kulikov, Ye.Ye., Isayeva A. S., Rotkina A. S., et al. (2007). Diversity and dynamics of bacteriophages in horse feces. *Microbiology* (Mikrobiologiya), 76(2), 271-278 [in Russian].

[16] Breitbart, M., Hewson, I., Felts, B., et al. (2003). Metagenomic analyses of an uncultured viral community from human feces. *J. Bacteriol.*, 185 (20), 6220-6223.

[17] Breitbart, M., Haynes, M., Kelley, S., et al. (2008). Viral diversity and dynamics in an infant gut. *Res. Microbiol.*,159, 367-373.

[18] Cornax, R., Morinigo, M. A., Gonzalez-Jaen, F., et al. (1994). Bacteriophages presence in human faeces of healthy subjects and patients with gastrointestinal disturbances. *Zentralbl Bakteriol.*, 281 (2), 214-224.

[19] Finkbeiner, S. R., Allred, A. F., Tarr, P. I., et al. (2008). Metagenomic analysis of human diarrhea: viral detection and discovery. *PLoS Pathogens*, 4 (2), e1000011.

[20] Lepage, P., Colombet, J., Marteau, P., et al. (2008). Dysbiosis in inflammatory bowel disease: a role for bacteriophages? *Gut*, 57 (3), 424-425.

[21] Letarov, A. and Kulikov, E. (2009). The bacteriophages in human- and animal body-associated microbial communities. *J. Applied Microbiol.,* 107, 1-13.

[22] Klumpp, J., Dorscht, J., Lurz, R., et al. (2008). The terminally redundant, nonpermuted genome of Listeria bacteriophage A511: a model for the SPO1-like Myoviruses of gram-positive bacteria. *J. Bacteriol.,* 190 (17), 5753-5765.

[23] Lusiak-Szelachowska, M., Weber-Dabrowska, B., and Gorski, A. (2006). The presence of bacteriophages in human faces and their potential importance. *Pol. Merkur Lekarski.*, 21(124), 381-383.

[24] Wommack, K. E. and Colwell, R. R. (2000). Virioplankton: viruses in aquatic ecosystems. *Microbiol. Mol. Biol. Rev.*, 64 (1), 69-114.

[25] Abedon, T. S., Kuhl, J. S., Blasdell, G. B., and Kutter, M. E. (2011). Phage treatment of human infections. *Bacteriophage*, 1 (2), 66-85.

[26] Mai, V., Ukhanova, M., Visone, L., et al. (2010). Bacteriophage administration reduces the concentration of Listeria monocytogenes in the gastrointestinal tract and its translocation to spleen and liver in experimentally infected mice. *Int. J. Microbiol.*, 2010, 624234

[27] Sulakvelidze, A. (2011). Bacteriophage: A new journal for the most ubiquitous organisms on Earth. *Bacteriophage*, 1,1-2.

[28] Pogorelskaya, S. A., Belova, T. N., Simonovich, V. V., et al. (1968). Experience in the wide use of bacteriophages for prophylaxis of acute intestinal diseases. *Zh. Mikrobiol. Epidemiol. Immunobiol.*, 4, 34-37 [in Russian].

[29] Merabishvili, M., Pirnay, J. P, Verbeken, G., et al. (2009). Quality-controlled small-scale production of a well-defined bacteriophage cocktail for use in human clinical trials. *PLoS One*, 4, e4944.

[30] Boytsov, A. G., Porin, A. A., Lastovka, O. N., et al. (2006). *Bacteriophages*. Ivanov, V. P. (Ed.). Spb.: Sankt-Peterburgskaya meditsinskaya akademiya [in Russian].

[31] Vorob'ev, A. A., Pak, S. G., Savitskaya, K. I., et al. (1998). *Dysbacteriosis in children*. M.: Moskovskaya meditsinskaya akademiya [in Russian].

[32] Krylov, V. N. (2001). Phage therapy in terms of bacteriophage genetics. *Genetika*, 37(7), 869-887 [in Russian].

[33] Fortuna, W., Międzybrodzki, R., Weber-Dąbrowska, B., and Górski, A. (2008). Bacteriophage therapy in children: facts and prospects. *Med. Sci. Monit.,* 14(8), 126-132.

[34] Borysowski, J., Dąbrowska, K., Ohams, M. et al. The response of the immune system to phage: potential associations with phage therapy. In: *Abstract book,* conference: "Bacteriophages and Probiotics – Alternatives to Antibiotics" dedicated to the 120th birth anniversary of professor George Eliava, July 1-4, 2012, Tbilisi, Georgia, 33.

[35] Loc-Carrillo, C. and Abedon, T. S. (2011). Pros and cons of phage therapy. *Bacteriophage*,1 (2), 111-114.

[36] European Food Safety Authority (2012). Scientific Opinion on the evaluation of the safety and efficacy of ListexTM P100 for the removal of Listeria monocytogenessurface contamination of raw fish. EFSA Panel on Biological Hazards (BIOHAZ) European Food Safety Authority (EFSA), Parma, Italy, *EFSA Journal*, 10(3), 2615.

[37] US EPA (2005). *Biopesticides registration action document* (Xanthomonas campestris pv. vesicatoria and Pseudomonas syringae pv. tomato specific Bacteriophages) (Chemical PC Codes 006449 and 006521).

[38] US FDA (2006). Department of health and Human services Food and Drug Administration, 21 CFR Part 172 [Docket No. 2002F–0316 (formerly 02F-0316)], Food Additives Permitted for Direct Addition to Food for Human Consumption; Bacteriophage Preparation. Agency: Food and Drug Administration, HHS. Action: Final rule. *Federal Register* Vol. 71, No. 160 (Friday, August 18, 2006) *Rules and Regulations*, P.47729-47732.

[39] Boratyński, J., Syper, D., Weber-Dabrowska, B., et al. (2004). Preparation of endotoxin-free bacteriophages. *Cellular Molec. Biology Letters*, 9, 253-259.

[40] Carlton, R. M., Noordman, W. H., Biswas, B., et al. (2005). Bacteriophage P100 for control of Listeria monocytogenes in foods: genome sequence, bioinformatic analyses, oral toxicity study, and application. *Regulatory Toxicol. Pharmacol.*, 43, 301-312.

[41] Bondarenko, V. M. (2011). Clinical efficiency and the ways of rational application of bacteriophage treatment in clinical practice. *Prilozheniye k Zhurnalu infektologii,* 3(3), 15-19 [In Russian].

[42] Zakharova, Yu. A., Nikolayeva, A. M. and Padrul, M. M. (2010). Bacteriophage treatment of urinary tract in pregnant women. *Biopreparaty*, 2(38), 14-17 [In Russian].

[43] Zheludeva, I. V., Zhilenkov, Ye. L., Maksimovskaya, L. N., et al. (2002). Bacteriophage treatment of inflammatory paradontosis diseases. *Parodontologiya*, 1(2), 46-50 [In Russian].

[44] Russian Ministry of Health standard: «Intestinal dysbacteriosis. Protocol for the treatment of patients». (2003). [Otraslevoy standart. Sistema standartizatsii v zdravookhranenii rossiyskoy federatsii. Disbakterioz kishechnika. Protokol vedeniya bolnykh. *Utverzhdeno prikazom Minzdrava Rossii ot «9» iyunya 2003* g. №231] [In Russian].

[45] Babalova, E. G., Katsitadze, K. F., Sakvarelidze, L. A., et al. (1968). Prophylactic efficiency of dry dysentery bacteriophage. *Zh. Mikrobiol. Epidemiol. Immunobiol.*, 2, 143-145 [In Russian].

[46] Voroshilova, N. N., Alferova, E. V., Darbeyeva, O. S., et al. (2010). Investigation of clinical efficiency of polyvalent bacteriophage Enterobacter purified. *Biopreparaty*, 2(38), 31-33 [In Russian].

[47] Zhukov-Verezhnikov, N. N., Peremitina, L. D., Berillo, E. A., et al. (1978). Therapeutic effect of bacteriophage preparations in the complex treatment of suppurative surgical diseases. *Sov. Med.*, 12, 64-66 [In Russian].

[48] Litvinova, A. M., Chtetsova, V. M. and Kavtreva, I. G. (1978). Evaluation of efficacy of the use of E. coli-Proteus bacteriophage in intestinal dysbacteriosis in premature infants. *Voprosy okhrany materinstva i detstva,* 9, 42-44 [In Russian].

[49] Milyutina, L. N. (1990). Phage therapy of acute intestinal infections in children. In: Actual questions of intestinal infections. *Proceedings of the conference*, Tashkent: «FAN», 107-109 [In Russian].

[50] Nilova, L. Yu., Orishak, E. A. and Il'inskaya, S. L. (2007). Phage therapy of intestinal dysbiosis. *Vestnik SPbGMA*, 1 (Suppl.), 152-153 [In Russian].

[51] Shakhmardanov, M. Z., Zemskova L. N., Isaeva, N. P., and Luchshev, V. I. (1997). Polyvalent pyobacteriophage treatment of the acute intestinal infections. *Rus. Med. Journal*, 1, 54-55 [In Russian].

[52] Solodovnikov, Yu. P., Pavlova, L. I., Emelyanov, P. I., et al. (1970). Prophylactic application of dry polyvalent dysentery bacteriophage with pectin in children's pre-school institutions. *Zh. Mikrobiol. Epidemiol. Immunobiol.,* 5, 131-137 [In Russian].

[53] Markoishvili, K., Dzhavakishvili, N., Gordedzishvili, M., et al. (1999). PhageBioderm – new perspectives for the treatment of wounds and ulcers. *Eksperim klin meditsina*, 2, 83-84 [In Russian].

[54] Slopek, S., Weber-Dabrowska, B., Dabrowski, M., and Kucharewicz-Krukowska, A. (1987). Results of bacteriophage treatment of suppurative bacterial infections in the years 1981-1986. *Arch. Immunol. Ther. Exp.* (Warsz), 35, 569-583.

[55] Weber-Dabrowska, B., Mulczyk, M. and Gorski, A. (2003). Bacteriophages as an efficient therapy for antibioticoresistant septicemia in man. *Transplant. Proc.*, 35, 1385-1386.

[56] Cisło, M., Dabrowski, M., Weber-Dabrowska, B., and Woyton, A. (1987). Bacteriophage treatment of suppurative skin infections. *Arch. Immunol. Ther. Exp.* (Warsz), 35, 175-183.

[57] Abdul-Hassan, H. S., El-Tahan, K., Massoud, B., and Gomaa, R. (1990). Bacteriophage therapy of Pseudomonas burn wound sepsis. *Ann. Med. Burn Club*, 3, 262-264.

[58] Lang, G. P., Kehr, P., Mathevon, H., et al. (1979). Bacteriophage therapy of septic complications of orthopaedic surgery. *Rev. Chir. Orthop. Repar. Appar. Mot.*, 1, 33-37.

[59] Wright, A., Hawkins, C. H., Anggård, E. E., and Harper, D. R. (2009). A controlled clinical trial of a therapeutic bacteriophage preparation in chronic otitis due to antibiotic-resistant Pseudomonas aeruginosa; a preliminary report of efficacy. *Clinical Otolaryngology*, 34 (4), 349-357.

[60] OmniLytics Website: http://www.omnilytics.com/products/agriphage/ agriphage2.html

[61] Leverentz, B., Conway, W. S., Camp, M. J., et al. (2003). Biocontrol of Listeria monocytogenes on fresh-cut produce by treatment with lytic bacteriophages and a bacteriocin. *Applied and Envir. Microbiol.*, 69 (8), 4519-4526.

[62] Leverentz, B., Conway, W. S., Janisiewicz, W., Camp, M. J., et al. (2004). Optimizing concentration and timing of a phage spray application to reduce Listeria monocytogenes on honeydew melon tissue. *J. Food Protection,* 67 (8), 1682-1686.

[63] Munsch, P., Olivier, J. V. and Houdeau, G. (1991). Experimental control of bacterial blotch by bacteriophages. In: Maher, M. J. (Ed.), *Science and cultivation of edible fungi,* Rotterdam, The Netherlands: Balkema, 389-396.

[64] Balogh, B., Jones, J. B., Momol, M. T., et al. (2003). Improved efficacy of newly formulated bacteriophages for management of bacterial spot on tomato. *Plant Dis.*, 87, 949-954.

[65] Schnaebel, F. L., Fernando, W. G. D., Meyer, M. P., and Jones, A. I. (1999). Bacteriophages of Erwinia amylovara and their potential for biocontrol. In: Mornol, M. T. and Saygili, H. (Eds.), *Proc. of the 8th International Workshop on Fireblight International Society of Horticultural Science,* Leven, Belgium, 649-653.

[66] Randhawa, P. S. and Civerolo, E. L. (1986). Interaction of Xanthomonas campestris pv. pruni with pruni phage and epiphytic bacteria on detached peach leaves. *Phytopathology*, 76, 549-553.

[67] Pao, S., Randolph, S. P., Westbrook, E. W., and Shen, H. (2004). Use of bacteriophages to control Salmonella in experimentally contaminated sprout seeds. *J. Food Sci.,* 69, 127-130.

[68] Wu, J.-L., Lin, H.-M., Jan, L., et al. (1981). Biological control of fish bacterial pathogen, Aeromonashydrophila, by bacteriophage AH 1. *Fish Pathol.,* 15, 271-276.

[69] Park, S. C., Shimamura, I., Fukunaga M., et al. (2000). Isolation of bacteriophages specific to a fish pathogen, Pseudomonas plecoglossicida, as a candidate for disease control. *Appl. Environ. Microbiol.,* 66, 1416-1422.

[70] Sklar, I. B. and Joerger, R. D. (2001). Attempts to utilize bacteriophage to combat Salmonella enteric serovar Enteritidis infection in chickens. *J. Food Saf.*, 21, 15-29.

[71] Berchieri, A., Jr., Lovell, M. A., Barrow, P. A. (1991). The activity in the chicken food-borne tract of bacteriophages lytic for Salmonella Typhimurium. *Res. Microbiol.*, 142, 541-549.

[72] Huff, W. E., Huff, G. R., Rath, N. C., et al. (2002). Prevention of Escherichia coli respiratory infection in broiler chickens with bacteriophage (SPR02). *Poult. Sci.*, 81, 437-441.

[73] Bach, S. J., McAlister, T. A., Veira, D. M., et al. (2002). Transmission and control of Escherichia coli O157:H7 – a review. *Can. J. Anim. Sci.*, 82, 475-490.

[74] Gill, J. J., Pacan, J. C., Carson, M. E., et al. (2006). Efficacy and pharmacokinetics of bacteriophage therapy in treatment of subclinical Staphyloccocus aureus mastitis in lactating dairy cattle. *Antimicrob. Agents Chemother.* 50 (9), 2912-2918.

[75] Lee, N. and Harris, D. L. (2001). The effect of bacteriophage treatment to reduce the rapid dissemination of Salmonella in pigs. *Proc. Am. Assoc. Swine Vet.,* 32, 555-557.

[76] Micreos Website: http://www.micreosfoodsafety.com/en/profile-mission.aspx.

[77] Guenther, S., Huwyler D., Richard, S., and Loessner, J. M. (2009). Virulent bacteriophage for efficient biocontrol of Listeria monocytogenes in ready-to-eat foods. *Applied And Environmental Microbiology,* 75 (1), 93-100.

[78] Abuladze, T., Li, M., Menetrez, Y. M., et al. (2008). Bacteriophages reduce experimental contamination of hard surfaces, tomato, spinach, broccoli, and ground beef by Escherichia coli O157:H7. *Applied and Environmental Microbiology,* 74 (20), 6230-6238.

[79] Intralytix Website: http://www.intralytix.com/Intral_products.htm

[80] Hooton, S. P., Atterbury, R. J. and Connerton, I. F. (2011). Application of a bacteriophage cocktail to reduce Salmonella Typhimurium U288 contamination on pig skin. *Int. J. Food Microbiol.*, 151 (2), 157-163.

[81] Murthy, K., Allain, B. and Roman, J.-L. (2002). Phage therapy: an innovative approach to treat antibiotic-resistant bacterial infections. *Am. Assoc. Swine Vet.,* 217-220. Available from: www.biopharma.net/ images/stories/PDF/aasv_2002_fullpaper.pdf

In: Bacteriophages
Editors: Clark Denton and Richard J. Crosby

ISBN: 978-1-62618-513-5

Chapter 6

Use of Bacteriophages as Surrogate Indicators of Viruses in Water

Fatma Hmaied* and Sihem Jebri
Laboratoire de Microbiologie et Biologie Moléculaire, CNSTN,
Pôle technologique, Sidi Thabet, Tunisie

Abstract

There are many potential applications of bacteriophages as models to assess the behaviour and survival of enteric viruses in the environment, and as surrogates to assess the resistance of human viruses to water treatment and disinfection processes. To minimize pathogens health risk, members of the coliform bacterial group, including *Escherichia coli*, in drinking water are measured to determine whether water has been contaminated by fecal material. Nevertheless, viruses are more resistant to chlorine and other disinfectants used in water treatment than are coliforms.

Thus, fecal coliform numbers do not serve as adequate indicators of enteric virus contamination which may affect human health after consumption of products contaminated by reclaimed water or sludge.

The use of bacteriophages as indicators of the presence and behaviour of enteric bacteria and animal viruses has always been attractive because of the ease of detection and low cost associated with phage assay and for similarity to enteric viruses in terms of transport and survival characteristics. Somatic coliphages, F-specific RNA phages, and bacteriophages that infect *Bacteroides fragilis* have been proposed as potential indicators of water quality and as tracers for enteric viruses presence in water due to their viral nature and their similar size and structure.

* Corresponding author's email: hmaiedfatma@yahoo.com.

Introduction

Wastewater treatment plants residues contain a large amount of microorganisms that are pathogenic for humans (viruses, bacteria and parasites) which can be released in aquatic environments. Enteric viruses may be introduced in these environments through human activities such as leaking sewage and septic systems, urban runoff, agricultural runoff, and, in the case of estuarine and marine waters, sewage outfall and vessel wastewater discharge (WHO, 2004). These viruses have been related to emerging waterborne disease (CDC, 2010; EFSA, 2011). Over 100 types of pathogenic viruses are excreted in human and animal wastes. Then, enteric viruses are released in large quantities into the environment and resist to physicochemical conditions of environment. They persist despite the presence of environmental stressors such as heat, sunlight and UV exposure (Gerones et al. 1989, Gerba et al. 2002, Vasickova et al. 2010). Methods for enteric viruses detection are therefore needed for screening water for the presence of these viruses. The impracticability of monitoring for the presence of all viral pathogens suggest that the virological safety of a water cannot presently be reliably assessed by direct analyses for viruses. When a treatment process has become established as effective in removing viruses, it should be monitored on a routine basis. But, direct analyses of viruses are not, and will never be, suitable. The methods used for the detection of enteric viruses require specialized laboratories, are more expensive and time consuming than traditional tests for faecal indicator bacteria and may be beyond the reach of control agencies (IAWPRC Study group, 1991). In fact, the detection of waterborne viruses is more complex than detection of other microorganisms. Difficulties in concentrating samples and then in detecting the virus by cell culture or molecular biological mean are encountered (Percival et al. 2004). Moreover, most of human waterborne viruses could not be cultivable. Besides, genomic methods have the inconvenient that they do not allow distinguishing between infectious and non infectious viruses. These difficulties lead us to the concept of indicator microorganisms.

Various indicators have been introduced. An ideal surrogate should be similar in size and structure to the target virus, slightly more resistant to the physical or chemical treatments commercially employed, non-pathogenic, and easy to handle and propagate in a laboratory setting (Jofre 2007). Bacteriophages of enteric bacteria had been advocated as potential viral indicators. Phages share many properties with human enteric viruses, notably composition, morphology, and structure, which implies that they may reflect the behavior and resistance of viruses to treatment process much closer than bacterial indicators (Contreras-Coll et al. 2002; Skraber et al. 2004). Another valuable feature of phages is that they are detectable by simple and inexpensive techniques that yield results in a relatively short period of time. Also, phages do not constitute a human health risk.

1. General Characteristics of Bacteriophages

Bacteriophages are viruses that infect bacteria. They were independently discovered by Twort (1915) and D'Herelle (1917). Bacteriophages are extremely abundant in nature with a high diversity. There are more than 10^{31} phage particles in the biosphere, 3-10 times greater than the number of bacteria (Graham and Roger 2011; Weinbauer 2004). Moreover, viral ecologists demonstrated that the population of bacteriophages is highly dynamic, there are about 10^{23} phage infections per second worldwide (Graham and Roger 2011). Phages consist of a nucleic acid molecule enclosed by a protein coat called capsid and in some case a lipid envelope (Singleton and Sainsbury, 2002). Many phages also contain additional structures such as tails and spikes. Like all viruses, phages are obligate intracellular parasites, thus they can replicate only inside susceptible and metabolising host bacteria. The host-specificity of phages is mainly determined by receptor molecules on the surface of the bacteria. Phages attached to the receptors located in the cell wall are typically known as "somatic phages". Phages that infect their hosts through receptors on F pili are called "F-specific bacteriophages". Bacteriophages that infect an *E. coli* bacteria are named "coliphages". The present classification of phages, which has been adopted by the International Committee on Taxonomy of Viruses, is mostly based in phage morphology and characteristics of the nucleic acid (Fauquet et al., 2005). Seven Bacteriophage families hold phages of particular interest in water quality assessment (Table 1); (Joffre 2007).

Table 1. Bacteriophage families holding phages of particular interest in water quality assessment

Group (family)	Characteristics	Representative phages (host)
Myoviridae	dsDNA long contractile tail. capsids up to 100 nm isometric or elongated capsids	T2, T4 (*E. coli*)
Siphoviridae	dsDNA long non contractile tails capsids up to 60 nm isometric capsids	λ (*E. coli*) B40-8 (*Bacteroides fragilis*)
Podoviridae	dsDNA short tails capsids up to 65 nm isometric capsids	T7, P22 (*E. coli*)
Microviridae	ssDNA without tail capsids about 25-30 nm isometric capsids	ΦX174 (*E. coli*)
Leviviridae	ssRNA without tail capsids about 25 nm isometric capsid	f2, MS2, GA, Qβ, F1 (F-plasmid bearing bacteria)
Inoviridae	dsDNA capsids about 800x6 nm long and flexible rods	fd, M13 (F-plasmid bearing bacteria)
Tectiviridae	dDNA without tail capsids up to60 nm isometric capsids lipid membrane below the capsid	PRD1, PR722 (enterobacteria)

2. Use of Bacteriophages as Indicators of Pathogens in the Environment

Bacteriophages have so far been advocated as potential model microorganisms for various aspects of water quality control. Several studies report the use of bacteriophages as indicators and as tracers for presence of bacteria in wastewater treatment system (Lucena et al. 2003; Withey et al. 2005) to assess the efficiency of water treatment plants and disinfection processes. In fact, fecal pollution was recognised as being the cause of waterborne diseases allowing the development of strategies for water-quality management (WHO 2004). Many studies focused on phages of enteric bacteria such as *E. coli* (coliphages) and *Bacteroides fragilis* (IAWPRC 1991; Grabow 1995). Bacteriophages were also used for trackig sources of fecal pollution in water (Lucena et al. 1996; Jofre et al. 2011). In fact, identifying dominant sources of fecal pollution is critical for accurate assessment of public health risks.

3. Phages Used as Surrogate Indicators of Viruses

Bacteriophages have been proposed as valuable models or surrogates for enteric viruses as their morphology and survival characteristics resemble those of human enteric viruses that pose a health risk to water consumers if water has been contaminated with human faeces (NHMRC, 1996; Grabowet al., 2001; Harwood et al., 2005). The direct correlations between the presence of certain bacteriophages and intensity of fecal contamination were reported. Three main groups of bacteriophages have been considered as potential model microorganisms for various aspects of water quality assessment (Bitton, 2002; Hurst et al., 2002). Also, bacteriophages had been included recently in some water quality regulations in USA (USEPA 2006) Canada (Anonymous 2001) and Australia (Queensland Government 2005). Somatic coliphages (Kott et al., 1974; IAWPRC Study Group, 1991), F-specific RNA bacteriophages (Havelaar and Pot Hogeboom, 1988; IAWPRC Study Group, 1991) and bacteriophages infecting *Bacteroides fragilis* (Tartera and Jofre, 1987; IAWPRC Study Group, 1991) are the groups of bacteriophages proposed and thoroughly investigated as surrogate indicators of viruses in water.

3.1. Somatic Coliphages

Somatic coliphage refers to a wide spectrum of lytic members of the families *Myoviridae*, *Siphoviridae*, *Podoviridae* and *Microviridae* that infect *E. coli* and certain closely related members of the bacterial family *Enterobacteriacea* (Hayes, 1968). These specific viruses of *E.coli* are useful indicators of the potential presence of enteric viruses in water environments, and valuable models/ surrogates for assessing the behaviour of enteric viruses in water treatment and disinfection processes.

The methodology to detect somatic coliphages is very simple and results may be obtained in 4 hours. They attach to *E. coli* cell wall and may lyse the host cell in 20 to 30 min under optimal physiological conditions. They produce plaques of different size and morphology.

One of the drawbacks of somatic coliphages is their replication potential outside the gut. In fact, some of their hosts may multiply or metabolise in water environments to the extent that they support the replication of somatic coliphages. However, recent reports on the host range of phages infecting the hosts strains used in standardised methods (Muniesa et al., 2003), the densities of phages and bacteria needed for replication to occur and the factors that affect phage replication (Wiggins and Alexander, 1985; Muniesa and Jofre, 2004), suggest that the contribution of replication outside the gut to the numbers of somatic coliphages in water environments is negligible in most aquatic environments. Bacteriophages frequently used to study somatic coliphages resistance to environmental stressors or disinfection or behaviour in concentration methods are T-4, T-7 and ϕX147.

3.2. Male Specific F-RNA Coliphages

F-RNA bacteriophages have been most extensively studied due to their similarity (in size, shape, morphology and physiochemistry) to many pathogenic human enteric viruses, namely enteroviruses, caliciviruses, astroviruses and Hepatitis A and E virus (Jofre et al., 2011). F-specific bacteriophages, also called sexual coliphages, are those that infect bacteria through the sex pili, which are coded by the F-plasmid first detected in *E. coli* K-12. The F-plasmid is transferable to a wide range of gram-negative bacteria. F-specific RNA bacteriophages consist of a simple capsid of cubic symmetry of 21 to 30 nm in diameter and contain single-stranded RNA as the genome. They belong to the family *Leviviridae*. Their infectious process is inhibited by the presence of RNase in the assay medium, which can be used to distinguish between the F-specific RNA bacteriophages and the rod-shaped F-specific DNA bacteriophages of the family *Inoviridae*, which also infect the host cell through the sex pili. Additionally, F-RNA bacteriophages (*Leviviridae*) have been further subdivided into four genogroups, which are primarily associated with either human (groups II and III) or animal (groups I and IV) origins (Furuse 1987). Modern genomic analysis is further identifying intermediate and possibly additional genogroups (Vinjé et al. 2004). The temperature conditioned production of sexual pili, necessary for phage replication. Then, F plasmid is not synthesized below 32 °C. Besides, Jofre et al. (2011) suggested that the environmental multiplication of these phages is unlikely because replication does not occur in nutrient-poor environments and requires a minimum host density of 10^4 colony forming units (cfu) per mL. Therefore the probability of these phages replicating in the environment is very small. A study on the effect of host cell and phages numbers, nutrition and competition from insusceptible cells, indicates that replication outside the gut is very improbable and therefore the influence of replication outside the gut in the numbers of these phages in the environment is negligible (Woody and Cliver, 1997).

The use of these phages as fecal pollution indicators has been proposed, because of their inability to replicate in the water ecosystem. A review of literature indicates that the most likely coliphages for use as a water quality indicator is the F-specific RNA coliphages. The group comprises viruses similar in size, shape, and genetic makeup, to human enteric viruses, which are responsible for most of the waterborne diseases. It represents viruses, which are more stable than human enteroviruses in environmental waters and more resistant to disinfections. Its concentration found in environmental waters has been reported to correlate with sewage contamination (Havelaar and Pot-Hogeboom, 1988). A standardized method for the detection and enumeration of this phage is now available; results can be obtained within 12 hours.

Theoretically, the use of a male (F + or Hfr) host strain would result in the detection of both somatic and F-specific phages (IAWPRC). Together, somatic and F-specific bacteriophages counts in water samples are usually designated as "total coliphage count". Some bacterial strains can be used to enumerate both simultaneously (Guzmán et al., 2008; Jebri et al. 2011). Their enumeration may be a good alternative for determination of viral contamination in poorly contaminated waters such as groundwater and drinking water or in double disinfection water treatments (Lucena and Jofre, 2010).

3.2. Phages of *Bacteroides*

The genome of the few *Bacteroides*-infecting phages that have been studied consists of double-stranded DNA, corresponding to that of *Siphoviridae* with flexible and frequently curved and curly tails. They attach to the cell wall of the host bacteria and may lyse the host cell in 30-40 minutes under optimal conditions (Kory and Booth 1986; Klieve et al. 1991; Puig and Gironés 1999; Queralt et al., 2003; Hawkins et al. 2008).

The great majority of *Bacteroides fragilis* phages have a narrow host range (Tartera and Jofre, 1978). Concurrently, strains of *Bacteroides fragilis* differ in the numbers of phages that they recover from sewage or sewage polluted waters (Puig et al., 1999).

Interest in technology and expertise for distinction between fecal pollution of human and animal origin based on microbiological methods is growing, and the term "microbial source tracking" (MST) became established (Scott et al. 2002). *Bacteroides fragilis* became interesting in this field as an (MST). In fact, they differ in their capability to discriminate human from animal faecal contamination. Thus, strain RYC2056 and VPI3625 of *Bacteroides fragilis* detect phages both in human and nonhuman fecal wastes (Kator and Rhodes 1992; Puig et al. 1999; Blanch et al. 2006), whereas *Bacteroides fragilis* HSP40 (Tartera and Jofre 1987; Tartera et al. 1989; Grabow et al. 1995), *Bacteroides thetaiotaomicron* GA17 (Payan et al. 2005; Blanch et al. 2006), and B. fragilis GB124 (Payan et al. 2005; Ebdon et al. 2007) detect phages mostly in human fecal wastes (much lower numbers if any in nonhuman fecal wastes).

Strains of *Bacteroides* spp. differ in the numbers of phages that they recover in different areas in the world, whereas RYC2056 detects similar numbers of phages in sewage around the globe (Puig et al., 1999; Lucena et al., 2003), strain HSP40 shows geographic differences, and whereas detecting good numbers, from 10^3 to 10^4 PFU 100 mL^{-1}, in the Mediterranean area (Tartera et al., 1989; Armon, 1993) and South Africa (Grabow et al., 1993), it fails to detect significant numbers in Northern Europe (Puig et al., 1999) and the USA (Chung et al., 1998). Strain VPI3625 tested in USA seems to behave similarly to strain RYC2056 (Chung et al., 1998). A strain of *Bacteroides tethaiotaomicron* named GA17, has recently been reported to detect nearby 10^5 PFU 100 mL^{-1} of municipal and hospital wastewater in the Mediterranean area, but it shows geographic differences (Payán et al., 2005). However an easy method to isolate *Bacteroides* hosts convenient for a given geographical area has been described (Payán et al., 2005)

Bacteriophages infecting *Bacteroides fragilis* have not been reported to replicate outside the gut likely because of the special requirements, anaerobiosis and nutrients, of the host strain to support phage replication (Tartera and Jofre, 1987). They have been also reported to be resistant to unfavourable conditions. Their resistance to stressors such as heat, UV radiation, and chemical disinfectants surpasses that of the traditional bacterial indicators and is similar to that of the most resistant viruses and other groups of bacteriophages (Tartera et al., 1988; Tartera and Jofre, 1989; Jofre et al., 1989). These findings support their value as models/surrogates for the potential presence of enteric viruses in water environments, and as models/surrogates to assess the survival of enteric viruses to water treatment and disinfection processes.

4. Methods for Phages Detection

Bacteriophages are considered non-pathogenic to humans, and can be readily cultured and enumerated in the laboratory. The methods to recover coliphages from environmental waters is relatively simple and within the capability and resources of most water quality laboratories (Debartolomeis and Cabelli, 1991). Numbers of phages are determined by direct quantitative plaque assay.

Bacteriophages can replicate only inside susceptible and metabolising host bacteria and can infect only certain bacteria. The host-specificity of phages is mainly determined by receptor molecules on the surface of the bacteria. Phage receptors had been described in different parts of bacteria (capsule, cell wall, flagella and pili). Phages attached to the receptors located in the cell wall are typically known as somatic phages. (Jofre, 2007).

ISO Methods for somatic coliphages (ISO 2000) and F-specific RNA Bacteriophages (ISO 1995) are available. Both include a double layer agar procedure for quantification and a presence–absence test that can be adapted to the Most Probable Number procedure MPN. Methods approved by the US EPA include the two-step enrichment procedure (US EPA 2001a) and the single agar layer assay (US EPA 2001b). The single agar layer and the double agar layer methods are plaque assay methods used to enumerate coliphages in volumes up to 100 mL (Grabow and Coubrough 1986). The two step enrichment procedure involves a liquid culture enrichment of 1 mL samples. The presence of phages in a given volume of sample can also be determined by the qualitative presence–absence enrichment test. Enrichment is accomplished by adding host bacterium and nutrients to a sample and then incubating under conditions that permit infection of the bacteria and replication of the phages present in the sample. The number of phages increases to the point where they are readily detectable by plaque assay or spot test on a lawn of the host strain (Adams 1959). This qualitative test can be adapted to "quantal" methods such as the MPN test using multiple sample volumes (Kott et al. 1974; Sobsey et al. 2004). Rapid methods are also being developed using 2 to 5 hours enrichment steps followed by latex agglutination (Love and Sobsey 2007) or bioluminescence relating to host lysis (Guzmán et al. 2009). In ISO methods, there are some fast but less controlled alternatives that can facilitate the use in laboratories with scarce facilities.

F-specific RNA bacteriophages infectious process is inhibited by the presence of RNase in the assay medium, which can be used to distinguish between the F-specific RNA bacteriophages and the rod-shaped F-specific DNA bacteriophages of the family *Inoviridae*, which also infect the host cell, through the sex pili. Molecular approaches to F-specific bacteriophages detection have also been developed for detecting phages in hours, without the need for culture. Reverse transcription-polymerase chain reaction (RT-PCR) assays have been developed for MS2, a prototypical F-RNA bacteriophage (O'Connell et al. 2006). Multiplexed RT-PCR assays have also been published to simultaneously detect F-RNA and F-DNA coliphages, coupled with a reverse line blot hybridization technique to genotype male-specific coliphages (Vinjé et al. 2004). More recently, multiplex RT-PCR assays have been introduced to distinguish the four genogroups of F-RNA coliphages (Ogorzaly and Gantzer 2006; Kirs and Smith 2007; Friedman et al. 2009).

Plaque assays for *Bacteroides fragilis* phages are more complicated, expensive, labour-intensive and time-consuming than those for somatic and F-RNA coliphages. A standardized method for the detection and enumeration of *Bacteroides fragilis* phages is now available (ISO, 2001). It is slightly more complex and costly than methods for detecting the other groups. Results can be obtained in 18 hours. Rather complex growth media supplemented with antibiotics are required, and plates have to be incubated under strictly anaerobic conditions. Theoretically, waste waters may be expected to contain higher levels of *Bacteroides fragilis* phages than those on record. The reason for the discrepancy may largely be due to failure of maintaining sufficiently anaerobic conditions during the performance of plaque assays. The phage mostly used to study phage resistance to environmental stressors or disinfection or behaviour in concentration methods is B40-8.

Methods described for concentrating F-RNA bacteriophages are also suitable to concentrate phages infecting *Bacteroides fragilis* (Mendez et al. 2004). Molecular methods for the detection of bacteriophages infecting *Bacteroides fragilis* are feasible (Puig et al. 2000), though further investigation about these procedures is needed. The availability of the sequence of the genome of a *B. fragilis* infecting phage will certainly facilitate this need (Hawkins et al. 2008).

5. Limitations

Somatic coliphages are not usually considered good fecal indicators because some of their hosts are unlikely to be of fecal origin (Hsu et al. 1996), and some of these phage are able to multiply in waters not subjected to fecal pollution (Gerba, 2006). But it seems that the contribution of replication outside the gut to the numbers of somatic coliphages found in water samples is negligible (Jofre, 2009). Moreover, they are not predictive indicators of virus presence or absence in groundwater (Payment and Locas, 2011), though some somatic phage such as T-4, T-7, ΦX174, and PRD-1 have proven useful as viral surrogates of fate and transport in laboratory investigations, pilot trials, and validation testing (WHO, 2004; Lucena and Jofre, 2010).

The persistence of F-specific RNA bacteriophages in surface waters, mainly in warm climates, is low (Chung and Sobsey, 1993; Moce´-Llivina et al., 2005), as is their resistance to some inactivating treatments, such as those based on heating or liming (Feng et al., 2003; Moce´-Llivina et al., 2003). For some purposes, such as determination of viral contamination in minimally contaminated waters like groundwater and drinking water (Federal Register, 2006), or reclamation treatments including two disinfecting treatments (for example, UV irradiation and chlorination) (Costa´n-Longares et al., 2008), total coliphages might be a good alternative (Jebri et al. 2011).

Some bacterial strains of *Bacteroides* detect phages with an unequivocal origin in the human gut. They are more resistant than the other groups to most natural stressors and water treatments. The method for detection requires cultivation of the host in anaerobiosis. Their low numbers in water and the need for different hosts for different geographic areas are at present the main weak points. In contrast, they seem very well suited as source tracers of fecal contaminants (Blanch et al., 2006; Ebdon et al., 2007).

General Conclusion

These data showed that Bacteriophages infecting enteric bacteria could be used as surrogate indicators of viruses in water due to their numbers in fecally contaminated and their similar properties, mainly their resistance to environmental stressors. Moreover, Phages are easy to handle and propagate in a laboratory setting.

Some regulatory agencies included bacteriophages in their detection policies. The U.S. Environmental Protection Agency has included coliphages in the Ground Water Rule (Federal Register, 2006); the Canadian Province of Québec has included coliphages in the regulation about drinking water quality (Gouvernement du Québec, 2001); and a few U.S. states have regulations regarding the quality required to reclaim water for certain uses (U.S. Environmental Protection Agency, 2004). Also, the U.S. Environmental Protection Agency is using bacteriophages in the validation and evaluation of membrane integrity (U.S. Environmental Protection Agency, 2006) and for validation of UV lamps to be used for water disinfection (U.S. Environmental Protection Agency, 2003).

References

Adams, M. H. (1959). Bacteriophages. New York: Interscience Publishers, Inc. Sobsey, M. D., M. V. Yates, F.-C. Hsu, G. Lovelace, D. Battigelli, A. Margolin, S. D. Pillai, and N. Nwachuku. Development and evaluation of methods to detect coliphages in large volumes of water. *Water Sci. Technol. 50*, 211–217.

Anonymous (2001). Loi sur la qualité de l'environement : règlement sur la qualité de l'eau potable c. Q-2, r.18.1.1. Gazette Officielle du Québec, *24*, 3561.

Armon, R. (1993). Bacteriophages monitoring in drinking water: do they fulfil the index or indicator function. *Water Science and Technology*, *27*, 463-467.

Blanch, A. R., Belanche-Munoz, L. & Bonjoch, X. et al. (2006). Integrated analysis of established and novel microbial and chemical methods for microbial source tracking. *Appl Environ Microbiol*, *72*, 5915–5926

CDC (2010). Surveillance for foodborne disease outbreaks e United States, 2007. *MMWR*, *59* (31), 973e979.

Chung, H. & Sobsey, M. D. (1993). Comparative survival of indicator viruses and enteric viruses in seawater and sediment. *Water Science and Technology*, *27*, 425-429.

Chung, H., Jaykus, L-A., Lovelance, G. & Sobsey, M. D. (1998). Bacteriophages and bacteria as indicators of enteric viruses in oysters and their harvest waters. *Water Science and Technology*, *38*, 37-44.

Contreras-Coll, N., Lucena, F., Mooijman, K., Havelaar, A., Pierz, V., Boque, M., Gawler, A., Holler, C., Lambiri, M. & Mirolo, G. et al., (2002). Occurrence and levels of indicator bacteriophages in bathing waters throughout Europe. *Water Res. 36*, 4963–4974.

Costa´n-Longares, A., Montemayor, M., Paya´n, A., Mendez, J., Jofre, J., Mujeriego, R. & Lucena, F. (2008). Microbial indicators and pathogens: removal, relationships and predictive capabilities in water reclamation facilities. *Water Res. 42*, 4439–4448.

DeBartolomeis, J. & Cabelli, V. J. (1991). Evaluation of an Escherichia coli host strain for enumeration of F male-specific bacteriophages. *Applied and Environmental Microbiology, 57*, 1301-1305.

D'Herelle, F. (1917). Sur un microbe invisible antagonistic des bacilles dysenteriques. *C R Acad Sci Paris, 165*, 373–375

Ebdon, J., Muniesa, M. & Taylor, H. (2007). The application of a recently isolated strain of Bacteroides (GB-124) to identify human sources of faecal pollution in a temperate river catchment. *Water Res, 41*, 3683–3690

EFSA (2011). Scientific opinion on an update on the present knowledge on the occurrence and control of foodborne viruses. *EFSA J., 9* (7), 2190.

Fauquet, C., Mayo, M. A., Maniloff, J., Desselberger, U. & Ball, L. A. (2005). Virus Taxonomy: Eighth Report of the International Committee on Taxonomy of Viruses. Elsevier/Academic Press.

Federal Register. (2006). National primary drinking water regulations: Ground Water Rule, final rule. 40 CFR parts 9, 141, and 142. *Fed. Regist. 71*, 65574–65660.

Feng, Y. Y., Ong, S. L., Hu, J. Y., Tan, X. L. & Ng, W. J. (2003). Effects of pH and temperature on the survival of coliphages MS2 and Q_. *J. Ind. Microbiol. Biotechnol., 30*, 549–552.

Friedman, S. D., Cooper, E. M. & Casanova, L. et al. (2009). A reverse transcription-PCR assay to distinguish the four genogroups of male-specific (F+) RNA coliphages. *J Virol Methods, 159*, 47–52.

Furuse, K. (1987). Distribution of coliphages in the environment: general considerations In: Goyal SM, Gerba CP, Bitton (1987), G (ed.) Phage Ecology. Wiley, New York

Gerba, C. P., Gramos, D. M. & Nwachuku, N. (2002). Comparative inactivation of enteroviruses and adenoviruses 2 by UV light. *Applied and Environmental Microbiology, 68*, 5167-5169.

Gerba, C. P. (2006). Bacteriophage as pollution indicators. In Calender R. (ed.) The Bacteriophages. Second edition. Oxford University Press. New York, USA.

Gironés, R., Jofre, J. & Bosch, A. (1989). Natural Inactivation of enteric viruses in seawater. *Journal of Environmental Quality, 18*, 34-39.

Grabow, WOK (2001). Bacteriophages: Update on application as models for viruses in water. *Water SA, 27*, 251–268

Grabow, WOK & Coubrough, P. (1986). Practical direct plaque assay for coliphages in 100-ml samples of drinking water. *Appl Environ Microbiol, 52*, 430–433

Grabow, W. O. K., Neubrech, T. E., Holtzhausen, C. S. & Jofre, J. (1995). Bacteroides fragilis and Escherichia coli bacteriophages. Excretion by humans and animals. *Water Science and Technology, 31*, 223-230.

Hatfull, Graham F. & Hendrix, Roger W. (2011). Bacteriophages and their genomes. *Current Opinion in Virology 2011, 1*, 298–303

Guzmán, C., Costan-Longares, A., Lucena, F. & Jofre, J. (2009). Detection of somatic coliphages through a bioluminescence assay measuring phage mediated release of adenylate kinase and adenosine 5'-triphosphate. *J Virol Methods*, *161*, 107–113

Guzmán, C., Mocé-Levina, L., Lucena F. & Jofre J. (2008). Evaluation of Escherichia coli host strain (CB390) for simultaneous detection of somatic and F-specific coliphages. *Applied Environmental Microbiology*, *74*, 531-534.

Harwood, V. J., Levine, A. D., Scott, T. M., Chivukula, V., Lukasik, J., Farrah, S. R. & Rose, J. B. (2005). Validity of the indicator organism paradigm for pathogen reduction in reclaimed water and public health protection. *Appl. Env. Micro.*, *71*, 3163-3170.

Havelaar, A. H. & Pot-Hogeboom, W. M. (1988). F-specific RNA bacteriophages as model viruses in water hygiene: ecological aspects. *Water Sci. Technol. 20*, 399–407.

Hawkins, S. A., Layton, A. C. & Ripp, S. et al. (2008). Genome sequence of the Bacteroides fragilis ATCC 51477-B1. *Virol J*, *5*, 97–101

Hawkins, S. A., Layton, A. C. & Ripp, S. et al. (2008). Genome sequence of the Bacteroides fragilis ATCC 51477-B1. *Virol J*, *5*, 97–101

HAYES, W. (1968). The Genetics of Bacteria and their Viruses (2nd edn.) Blackwell Scientific Publications, Oxford.

Hsu, F. -C., Chang, A., Amante, A., Shieh, Y. -S. C., Wait, D. & Sobsey, M. C. (1996). Distinguishing human from animal feacal contamination in water by typing male specific RNA coliphages. *AWWA Water Technology Conference 1996.*

Hurst, C. J., Crawford, R. L., Knudsen, G. R., McInerney, M. J. & Stetzenbach, L. D. (2002). Manual of Environmental Microbiology., 2th *Ed. ASM Press*, Washington, DC.

IAWPRC (1991). Bacteriophages as model viruses in water quality control. *Water Res*, *25*, 529–545.

Percival, S., Chalmers, R., Embrey, M., Hunter, P., Sellwood, J., Wyn-Jones, P. (2004). Methods for the detection. *Microbiology of Waterborne Diseases*, 349-377.

ISO. (1995). ISO 10705-1: Water quality. Detection and enumeration of bacteriophages -part 1: Enumeration of F-specific RNA bacteriophages. Geneva. *International Organisation for Standardisation.*

ISO. (2000). ISO 10705-2: Water quality. Detection and enumeration of bacteriophages -part 2: Enumeration of somatic coliphages. Geneva, *International Organisation for Standardisation.*

ISO. (2001). ISO 10705-4: Water quality. Detection and enumeration of bacteriophages -part 4: Enumeration of bacteriophages infecting Bacteroides fragilis. Geneva. *International Organisation for Standardisation.*

ISO. (2003). ISO 10705-3. Water Quality. Enumeration and Detection of Bacteriophages. Part 3. Validation of Methods for Concentration of Bacteriophages from Water. Geneve. *International Standardisation Organization.*

Jebri, S., Jofre, J., Barkallah, I., Saidi, M. & Hmaied F. (2012). Presence and fate of coliphages and enteric viruses in three wastewater treatment plants effluents and activated sludge from Tunisia. *Environ Sci Pollut Res*, *19*, 2195–2201.

Jofre , J. (2007). «*Indicators of Waterborne Enteric Viruses*». In Bosch, A., Human Viruses in Water (Series Perspectives in Medical Virology). Elsevier. London.

Jofre, J., Stewart, J. R. & Grabow, W. G. (2011). Phage Methods, in Hagedorn C. Blanch A.R and Harwood V.J. (eds.), Microbial Source Tracking: Methods, Applications, and Case Studies, 1st edition. Springer, NY, USA.

Jofre, J. (2002). Bacteriophages as indicators, 354– 363. In G. Bitton (ed.), *Encyclopedia of Environmental Microbiology*, vol. *1*. John Wiley & Sons, Inc., New York, NY.

Jofre, J., Blasi, M., Bosch, A. & Lucena, F. (1989). Occurrence of bacteriophages infecting Bacteroides fragilis and other viruses in polluted marine sediments. *Water Science and Technology*, *21*, 15-19

Kator, H. & Rhodes, M. (1992). Evaluation of Bacteroides fragilis bacteriophages, a candidate human specific indicator of fecal contamination from shellfish growing waters. Final Report. Virginia Institute of Marine Sciences, College of Williams and Mary, Gloucester Point, Virginia, USA

Kirs, M. & Smith, D. C. (2007). Multiplex quantitative real-time reverse transcriptase PCR for F+-specific RNA coliphages: a method for use in microbial source tracking. *Appl Environ Microbiol*, *73*, 808–814

Klieve, A. V., Gregg, K. & Bauchop, T. (1991). Isolation and characterization of lytic phages from Bacteroides ruminicola ss brevis. *Curr Microbiol*, *23*, 183–187

Kory, M. M. & Booth, S. J. (1986). Characteristics of Bacteroides fragilis bacteriophages and comparison of their DNAs. *Curr Microbiol*, *14*, 199–203.

Kott, Y., Roze, N., Sperber, S. & Betzer, N. (1974). Bacteriophages as viral pollution indicators. *Water Res.*, *8*, 165–171.

Love, D. C. & Sobsey, M. D. (2007). Simple and rapid F+ coliphage culture, latex agglutination, and typing assay to detect and source track fecal contamination. *Appl Environ Microbiol*, *73*, 4110–4118

Lucena, F. & Jofre, J. (2010). Potential use of bacteriophages as indicators of water quality and wastewater treatment processes, in Sabour P. M and Griffiths M.W. (eds), Bacteriophages in the Control of Food and Waterborne Pathogen, ASM Press, New York.

Lucena, F., Mendez, X., Morón, A., Calderón, E., Campos, C., Guerrero, A., Cárdenas, M., Gantzer, C., Schwartzbrod, L., Skraber, S. & Jofre, J. (2003). Occurrence and densities of bacteriophages proposed as indicators and bacterial indicators in river waters from Europe and South America. *Journal of Applied Microbiology*, *94*, 808-815.

Lucena, F., Araujo, R. & Jofre, J. (1996). Usefulness of bacteriophages infecting Bacteroides fragilis as index microorganisms of remote faecal pollution. *Water Res*, *30*, 2812–2816

Lucena, F., Mendez, X. & Morón, A. et al. (2003). Occurrence and densities of bacteriophages proposed as indicators and bacterial indicators in river waters from Europe and South America. *J Appl Microbiol*, *94*, 808–815

Mendez, J., Audicana, A., Isern, A., Llaneza, J., Tarancon, M. L., Jofre, J. & Lucena, F. (2004). Standardised evaluation of the performance of a simple membrane filtration-elution method to concentrate bacteriophages from drinking water. *Journal of Virological Methods*, *117*, 19-25.

Mocé-Llivina, L., Muniesa, M., Pimenta-Vale, H., Lucena, F. & Jofre, J. (2003). Survival of bacterial indicator species and bacteriophages after thermal treatment of sludge and sewage. *Applied and Environmental Microbiology*, *69*, 1452- 1456.

Mocé-Llivina, L., Lucena, F. & Jofre, J. (2005). Enteroviruses and bacteriophages in bathing waters. *Applied and Environmental Microbiology*, *71*, 6838-6844.

Muniesa, M. & Jofre, J. (2004). Factors influencing the replication of somatic coliphages in the water environment. *Antoine van Leeuvenhoek*, *86*, 65-76.

Muniesa, M., Mocé-Llivina, L., Katayama, H. & Jofre, J. (2003). Bacterial host strains that support replication of somatic coliphages. *Antoine van Leeuvenhoek*, *83*, 305-315.

NHMRC (2001). In: Microbial indicators of water quality: an NHMRC, Discussion paper, National Health and Medical Research Council, Commonwealth of Australia.

O'Connell, K. P., Bucher, J. R., Anderson, P. E. et al. (2006). Real-time fluorogenic reverse transcription- PCR assays for detection of bacteriophage MS2. *Appl Environ Microbiol*, *72*, 478–483

Ogorzaly, L. & Gantzer, C. (2006). Development of real-time RT-PCR methods for specific detection of F-specific RNA bacteriophage genogroups: application to urban raw wastewater. *J Virol Methods*, *138*, 131–139

Payán, A., Ebdon, J., Taylor, H., Gantzer, C., Ottoson, J., Papageorgiou, G. T., Blanch, A. R., Lucena, F., Jofre, J. & Muniesa, M. (2005). Method for Isolation of Bacteroides bacteriophages host strains suitable for tracking sources of faecal pollution in water. *Applied and Environmental Microbiology*, *71*, 5659-5662.

Payment, P. & Locas A. (2011). Pathogens in water: value and limits of correlation with microbial indicators. *Ground Water*, *49*(1), 4-11.

Puig, M. & Gironés, R. (1999). Genomic structure of phage B40-8 of Bacteroides fragilis. *Microbiol* (UK), *145*, 1661–1670

Puig, A., Queralt, N., Jofre, J. & Araujo, R. (1999). Diversity of Bacteroides fragilis strains in their capacity to recover phages from human and animal wastes and from fecally polluted wastewater. *Applied and Environmental Microbiology*, *65*, 1772- 1776

Puig, M., Jofre, J. & Gironés, R. (2000). Detection of phage infecting Bacteroides fragilis HSP40 using a specific DNA probe. *Journal of Virological Methods*, *88*, 163-173.

Queralt, N., Jofre, J., Araujo, R. & Muniesa, M. (2003). Homogeneity of the morphological groups of bacteriophages infecting Bacteroides fragilis strain HSP40 and strain RYC2056. *Curr Microbiol*, *46*, 163–168

Scott, T. M., Rose, J. B., Jenkins, T. M., Farrah, S. R. & Lukasik, J. (2002). Microbial source tracking: current methodology and future directions. *Applied Environmental Microbiology*, *68*(12), 5796-5803.

Singleton, P. & Sainsbury, D. (2002). Dictionary of Microbiology 3rd ed. Chichester7 John Wiley and Sons.

Skraber, S., Gassilloud, B. & Gantzer, C. (2004). Comparison of coliforms and coliphages as tools for assessment of viral contamination in river water. *Appl. Environ. Microbiol. 70*, 3644–3649.

Tartera, C. & Jofre, J. (1987). Bacteriophages active against Bacteroides fragilis in sewage polluted waters. *Applied and Environmental Microbiology*, *53*, 1632-1637.

Tartera, C., Lucena, F. & Jofre, J. (1989). Human origin of Bacteroides fragilis bacteriophages present in the environment. *Applied and Environmental Microbiology*, *55*, 2696-2701.

Twort, F. W. (1915). An investigation on the nature of ultra-microscopic viruses. *Lancet*, *2*, 1241

U.S. Environmental Protection Agency. (2006). Environmental Technology Verification Report, Removal of Microbial Contaminants From Drinking Water, Koch Membrane Systems, Inc., HF-82-35-PMPW Ultrafiltration Membrane (PDF). EPA report no. 600-R-06-100. U.S. Environmental Protection Agency, Washington, DC.

Vasickova, P., Pavlik, I., Verani, M. & Carducci, A. (2010). Issues concerning survival of viruses on surfaces. *Food and Environmental Virology*, *2*, 24–34.

Vinjé, J., Oudejans, S. J. & Stewart, J. R. et al. (2004). Molecular detection and genotyping of male-specific coliphages by reverse transcription-PCR and reverse line blot hybridization. *Appl Environ Microbiol*, *70*, 5996–6004

Weinbauer, M. G. *Ecology of prokaryotic viruses*. FEMS Microbiol 2004 May; *28*(2), 127-81. Review.

WHO (2004). Water, sanitation and hygiene links to health: Facts and figures. World Health Organization, Geneva, Switzerland.

Wiggins, B. A. & Alexander, M. (1985). Minimum bacterial density for bacteriophages replication: implications for significance of bacteriophages in natural ecosystems. *Applied and Environmental Microbiology*, *49*, 19-23.

Withey, S., Cartmell, E., Avery, L. M. & Stephenson T. (2005). Bacteriophages—potential for application in wastewater treatment processes. *Science of the Total Environment*, *339*, 1– 18.

Woody, M. A. & Cliver, D. O. (1997). Replication of coliphage Q beta as affected by host cell number, nutrition, competition from insusceptible cells and non-FRNA coliphages. *J Appl Microbiol*, *82*, 431–440.

In: Bacteriophages
Editors: Clark Denton and Richard J. Crosby
ISBN: 978-1-62618-513-5

Chapter 7

Strategies for the Analysis and Engineering of Promoter Sequences for Bacteriophage RNA Polymerases

Shoji Ohuchi*
Department of Medical Genome Science, University of Tokyo,
Kashiwa-shi, Japan

Abstract

Because of their high transcription efficiency and easiness of manipulation, bacteriophage RNA polymerases are commonly used for *in vitro* transcription. In addition, their applicability for *in vivo* expression has also been demonstrated. Thus, bacteriophage RNA polymerases are promising components for artificial *in vitro* and prokaryotic expression systems. To widen the variety of independently regulated transcription machineries, the generation of bacteriophage RNA polymerase mutants with altered promoter specificities has been explored. Strict orthogonality of these mutants is required for practical applications, and therefore, the promoter specificities of the mutant RNA polymerases should be validated. For the analyses of promoter sequences, *in vivo* systems have been traditionally employed, but recently, several *in vitro* systems, by which huge numbers of promoter sequences can be analyzed with fewer complications, have been also developed. In this chapter, I review the progress in the development of experimental systems for the analyses and engineering of promoter sequences for bacteriophage RNA polymerases.

Introduction

Synthetic gene circuits have applications in the manipulation of biological systems as well as in the generation of simplified models for certain biological phenomena [1-3]. These

* Corresponding author: E-mail: sohuchi@ims.u-tokyo.ac.jp.

circuits are constructed by assembling natural, engineered, and/or *de novo* generated gene components. Because synthetic gene circuits have great potential not only for basic biological studies but also for industrial and medical applications, efforts are being made to design novel circuit types [1-3]. In prokaryotic cells and cell-free *in vitro* systems, regulation of gene expression is mainly achieved at the point of transcription initiation; therefore, transcription machineries are important components for synthetic gene circuits.

RNA polymerases from bacteriophages, including T7 and SP6 phages, are the most commonly used enzymes for the *in vitro* expression [4-7]. Furthermore, use of bacteriophage RNA polymerases for *in vivo* expression has also been demonstrated [8-12]. For example, tight regulation of transcription is achieved by using a combination of T7 RNA polymerase and its specific inhibitor, T7 lysozyme, and this expression system is widely used for the production of toxic proteins *in vivo* [13].

To widen the variety of independently regulated transcription machineries, the generation of mutants of bacteriophage RNA polymerases with altered promoter specificity has been explored. For example, several mutants of the T7 RNA polymerase/T7 promoter sequence have been designed [14, 15] on the basis of the crystal structure of the polymerase/promoter complex [16, 17] and comparative analysis of the amino acid/DNA sequences of the related polymerase/promoter pairs.

Strict orthogonality of the mutant RNA polymerase/promoter sequence pairs is required for practical applications; therefore, the promoter specificities of these mutant RNA polymerases should be validated. For the analyses of promoter specificities, *in vivo* systems, in which the reporter gene is expressed under the control of promoter sequences of interest, have commonly been used. Furthermore, several *in vitro* systems, wherein huge numbers of promoter sequences can be analyzed, have also been developed recently. Here, I review the progress in the development of experimental systems for the analysis and engineering of promoter sequences for bacteriophage RNA polymerases.

In Vivo Methodologies

Traditionally, *in vivo* gene fusion systems have been widely applied for the analysis and identification of promoter sequences [18-26]. In the system, promoter sequences of interest are fused to the coding sequence of a reporter gene, and the promoter activity is evaluated on the basis of the expression of the reporter protein. As a reporter, LacZ, a fragment of beta-galactosidase, has been the most commonly used protein because protein expression can be quantified easily by evaluating enzymatic activity using a chromogenic substrate [18-23]. In addition, use of this reporter enables viability-based selection when the cells are cultured in a media containing lactose as the sole carbon source.

Although LacZ-based systems can be used for the evaluation of the promoter activity, the results obtained are merely an average activity of a clonal population of cells harboring the same promoter sequence. Thus, homogeneity of the activity cannot be guaranteed in this system. For example, a heterogeneous population containing cells showing both strong promoter activity and weak activity cannot be distinguished from the homogeneous population of cells equally showing similar moderate activity. For practical application in metabolic engineering and synthetic biology field, the homogeneousness of activity is essential, and a system to evaluate the promoter activity at single cell resolution is desired.

Fluorescent proteins, such as green fluorescent protein (GFP) and its derivatives, are useful reporters to overcome the above problem [25-27]. Because of their strong fluorescence intensities, the expression level of the reporter in each cell can be quantitatively evaluated by using fluorescence activated cell sorter (FACS). For example, Alper et al. performed a FACS screening of mutants of a constitutive bacteriophage promoter, P_L-λ, using GFP as a reporter [26]. The authors identified several P_L-λ promoter derivatives with desired strength, as well as sustained homogeneousness, of the promoter activity. The identified promoter derivatives were examined for transgenic expression of lycopene biosynthetic pathway in *Escherichia coli*, and the result has strengthened the utility of the promoter engineering procedure.

In the typical reporter systems, promoter activity directly reflects the expression level of the reporter protein, and usually, the precise analyses of weak activities are difficult. Shin et al. have developed a unique reporter system to quantitatively evaluate weak promoter activity [28, 29]. In this system, the regulatory mechanism for plasmid replication at the ColE1 origin is utilized. The replication is mainly regulated by level of an RNA primer for the initiation of DNA synthesis [30] (Figure 1).

The RNA primer is transcribed as a long precursor, RNA II, which is encoded upstream of the replication origin. The transcribed RNA II forms a duplex with the template strand of the DNA, and following cleavage with RNase H generates the mature RNA primer. This process is regulated by a complementary strand to the 5'-regions of RNA II. This antisense transcript, RNA I, forms a duplex with RNA II and acts as a negative regulator for plasmid replication by inhibiting the RNase H-mediated primer maturation. The authors found that insertion of SP6 promoter sequence downstream of the replication origin severely decreased the plasmid copy number in the cells expressing SP6 RNA polymerase [31].

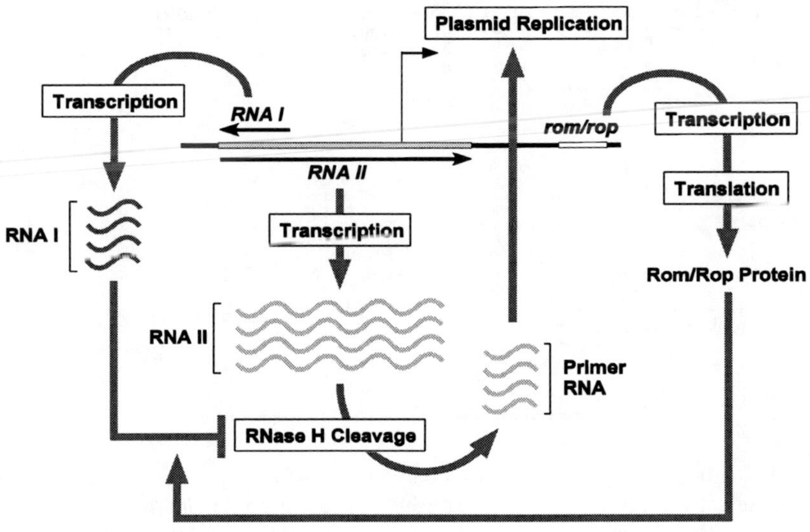

Figure 1. Regulatory mechanism for ColE1 plasmid replication. Transcribed RNA II is cleaved by RNase H to generate the mature RNA primer for the initiation of DNA synthesis. RNA I forms a duplex with RNA II and inhibits the RNase H-mediated cleavage. Thus, RNA I acts as a negative regulator for the plasmid replication. Rom/Rop protein facilitates the formation of RNA I/RNA II duplex.

This phenomenon was caused by the transcription of an RNA strand including the RNA I region. Detailed kinetics of the ColE1 plasmid replication has been well studied [32]; therefore, the transcription level of this RNA I-containing transcript can be calculated from the plasmid copy number. Thus, the authors quantitatively examined the activity of the promoter that was inserted downstream of the replication origin [28, 29] (Figure 2). In this system, strong and weak promoter activities lead to low and high levels of plasmid replication, respectively, and sensitive and accurate analyses of any weak activities can be achieved. By using this system, the authors examined the saturation mutagenesis analysis of SP6 promoter sequence *in vivo* [28, 29].

Esvelt et al. have developed a unique reporter system for the selection and evolution of genes of interest by employing the M13 bacteriophage [33]. In this system, the desired genes are encoded in the phage library, and expression of an active gene results in the production of the phage protein III (pIII). Because pIII is required for entry of the phage into the host cells [34, 35], only the phages carrying the active gene can propagate. Because of the quick life cycle and the large scale of the library that could be generated using the bacteriophage [35], the system is highly efficient. Combining this reporter system and continuous culture with induction of random mutagenesis, the authors succeeded in generating T7 RNA polymerase mutants that are capable of recognizing an altered promoter sequence, or transcribing RNAs with A or C as the initiating residues (rather than canonical G residue) [33]. Although the system has not been applied for the selection of promoter sequences, it would be very useful for the purpose.

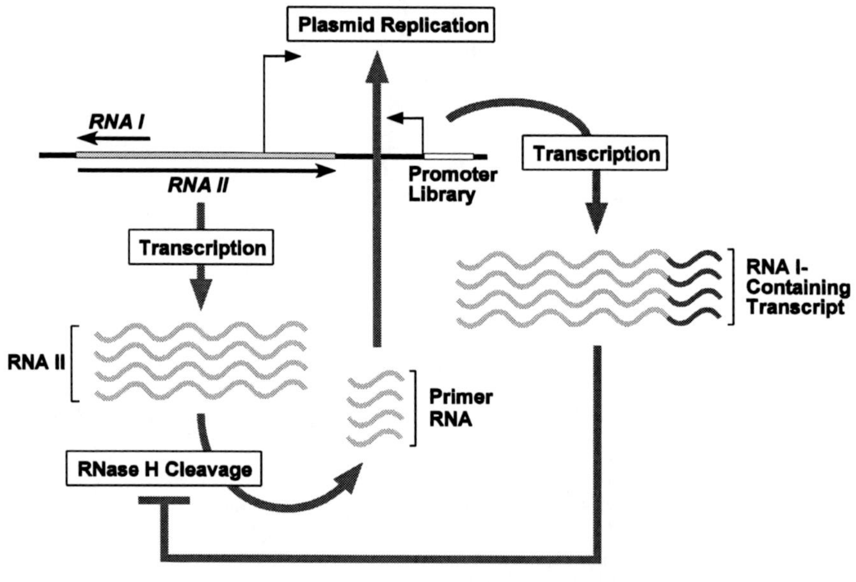

Figure 2. Evaluation of promoter activity based on ColE1 plasmid replication. The target promoter sequence is inserted at the downstream region of the replication origin, and its promoter activity leads to the generation of RNA I-containing transcript. Similar to natural RNA I, the transcript acts as a negative regulator for plasmid replication. Because the weaker promoter activity results in higher levels of the plasmid replication, the evaluation system achieves sensitive and accurate evaluation of the weak activity. Furthermore, the Rom/Rop protein is lacking in the system, and even the basal levels of the plasmid replication are higher than that of the natural ColE1 plasmid, resulting in higher sensitivity of the evaluation. The RNA polymerase is supplemented from a cotransformed plasmid.

In Vitro Compartmentalization-Based Methodologies

Although *in vivo* systems have been widely employed for the analysis and engineering of promoter sequences, they have several drawbacks including their smaller scale of library and limited experimental conditions (only intracellular conditions can be examined).

In vitro compartmentalization (IVC) is an *in vitro* system that can overcome these limitations [36-38]. In this system, small volumes of a reaction mixture are compartmentalized into micrometer-sized water-in-oil (W/O) emulsion, and numerous reaction compartments (up to 10^{10} compartments/mL) can be generated simply by vortexing the reaction mixture with surfactants and oil. When single molecules of template DNAs that are to be expressed by *in vitro* transcription (and translation) are compartmentalized into the W/O emulsion, each emulsion compartment can be treated as a tiny test tube for the screening of genes. Since the demonstration of IVC using a model experiment with DNA methylase [38], the system has been employed for the screening of proteins, peptides, and RNAs with desired binding and catalytic activities [36, 37]. By means of IVC, Davidson et al. have developed a screening system for promoter sequences by *in vitro* transcription of a bead-immobilized template DNA (Figure 3) [39]. In this system, a microbead which displays a single molecule of template DNA with the promoter sequence of interest and capture DNAs is compartmentalized with the RNA polymerase in W/O emulsion, and *in vitro* transcription is carried out. When the immobilized template DNA carries an active promoter sequence, the encoded RNA molecules are synthesized and captured onto the microbead via hybridization with the capture DNAs. After the reaction, the emulsion is disrupted, and the captured RNAs are fluorescently labeled by hybridization with the probe DNA, followed by signal amplification using antibodies. As the RNAs are captured only when the microbead displays the template DNA with an active promoter, the microbeads with desired promoter sequences can be selectively labeled and collected using FACS. Using this system, the authors examined the screening of the transcriptionally active template DNAs from a pool with six randomized sequences at the 5'-end of the coding region and demonstrated that the active sequences were enriched after three rounds of the screening.

Rescently, Kojima et al. have improved the above IVC-based system for screening promoter sequences (Figure 4) [40]. The first point of improvement is in the increased copy number of template DNA displayed onto the microbeads. Rather than simple immobilization of a single DNA molecule, they performed PCR amplification of a single DNA molecule in the W/O emulsion [41-43]. The PCR amplification, termed emulsion PCR, is performed using one primer immobilized onto the microbeads and another non-immobilized primer, and the amplified DNA fragments would then be immobilized onto the microbeads. Thus, when a single molecule of the template DNA is included in the single emulsion, emulsion PCR results in a microbead displaying multiple copies of a clonal population of the amplified DNAs. The higher copy number of the DNA is expected to enhance the sensitivity of the FACS screening. The second improvement is in the RNA capturing procedure that is based on ligation activity of an artificial ribozyme [44-46]. In contrast to Davidson's system, in which the RNAs are captured by hybridization, the RNAs are covalently coupled to the substrate RNA immobilized onto the microbeads. Because of its stable nature of the capturing, stringent washing of the microbeads is acceptable, and reduction of background signal would be achieved.

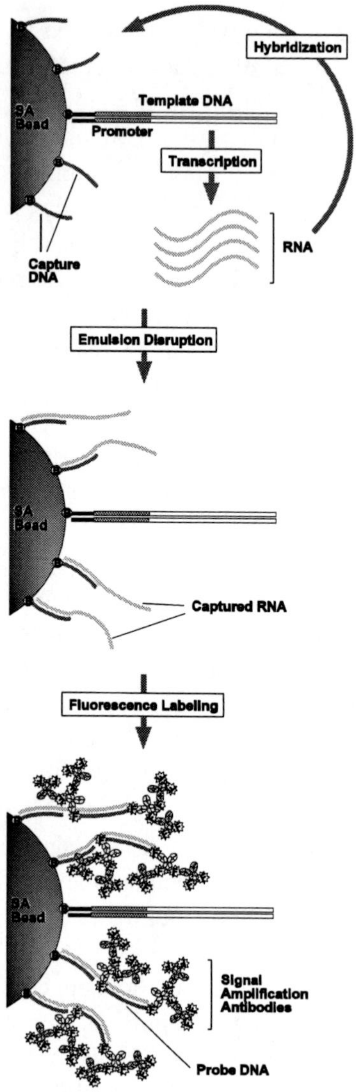

Figure 3. Promoter screening system employing *in vitro* transcription and hybridization in W/O emulsion. The template and the capture DNAs are immobilized onto the microbead via biotin-streptavidin interaction. The microbead is compartmentalized in W/O emulsion, and *in vitro* transcription is carried out. The synthesized RNAs are immobilized onto the microbead through hybridization with the capture DNAs. After the emulsion is disrupted, the captured RNAs are fluorescently labeled by hybridization with probe DNA followed by signal amplification with antibodies. Then, highly fluorescent microbeads, or the microbeads displaying the template DNA with an active promoter sequence, are selectively recovered using FACS. The recovered DNAs are amplified by PCR and utilized for the next round of the screening. “SA” means streptavidin. “B” and “F” indicate biotin and fluorescent moieties, respectively.

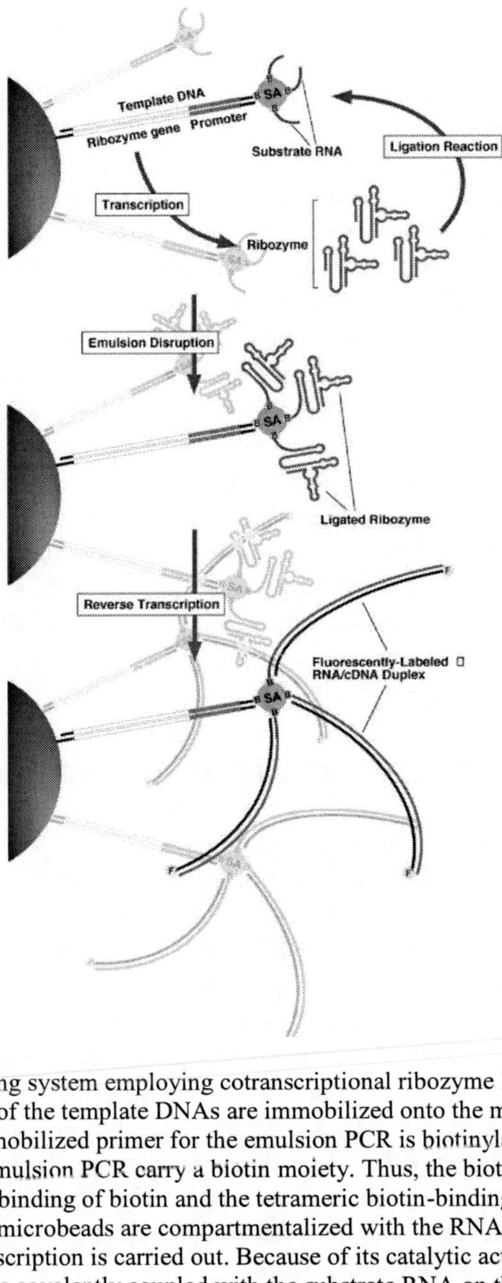

Figure 4. Promoter screening system employing cotranscriptional ribozyme ligation reaction in W/O emulsion. Multiple copies of the template DNAs are immobilized onto the microbead by emulsion PCR. Because the non-immobilized primer for the emulsion PCR is biotinylated, the immobilized template DNAs after the emulsion PCR carry a biotin moiety. Thus, the biotinylated substrate RNA can be displayed through tight binding of biotin and the tetrameric biotin-binding protein, streptavidin. The substrate RNA-displaying microbeads are compartmentalized with the RNA polymerase in W/O emulsion, and *in vitro* transcription is carried out. Because of its catalytic activity, the transcript, or an artificial ligase ribozyme, is covalently coupled with the substrate RNA on the microbeads. After disruption of the emulsion, the microbeads are subjected to reverse transcription with a fluorescently labeled primer. Through the process, the microbeads displaying the template DNA with active promoter sequence are selectively labeled, and can be recovered by using FACS. The recovered DNAs are amplified by canonical PCR and utilized for the next round of the screening. "SA" means streptavidin. "B" and "F" indicate biotin and fluorescent moieties, respectively.

By using the system, the authors performed a model experiment for the evaluation of transcription activities of pairs of T7 RNA polymerase/T7 promoter and their variations, and found that the fluorescence signal was detected only when the proper combination of promoter and RNA polymerase was employed. More importantly, the signal intensities were finely correlated to the reported transcription activity of the employed pair, demonstrating the applicability of the system for the evaluation of promoter activity. In addition, two rounds of the model screening using T7 RNA polymerase enriched the T7 promoter sequence to 70-fold in a mixture containing the SP6 promoter sequence.

Paul et al. have developed an alternative IVC-based system for screening promoter sequences [47]. In this system too, multiple copies of the template DNA are displayed onto microbeads via emulsion PCR, and *in vitro* coupled transcription/translation is performed in the W/O emulsion. Under the control of target promoter sequences, the fusion protein of *O*-6-alkylguaninalkyltranferase (SNAP) domain and GFP is expressed. Because the DNA is labeled with a suicide inhibitor of SNAP (benzylguanine) using a benzylguanine-conjugated primier during the emulsion PCR, the synthesized fusion protein is covalently coupled to the DNA by the enzymatic activity of SNAP [48]. Thus, after the emulsified reactions, the microbeads that contain an active promoter sequence display the fluorescent fusion protein and can be selectively recovered using FACS. Using this system, the authors performed the screening of sequence elements for efficient *in vitro* transcription from a library with random 10 nucleotide-long sequences (position -4 to -1 of the promoter region and the initial six bases of the coding region). After three rounds of screening, the authors successfully identified four sequence elements that showed the transcription activities higher than the original sequence including the wild-type T7 promoter.

In Vitro Selection Methodologies

Although IVC enables the screening of a large scale of libraries, *in vitro* selection procedures in bulk solution achieve far larger ones; more than 10^{14} varieties of sequences can be easily examined [49, 50]. In typical *in vitro* RNA selections, double-stranded DNA that consists of forward and reverse primer-binding sites, promoter sequence, and random or mutagenized sequence is used as a transcription template for RNA pool synthesis [49, 50]. The RNA pool is subjected to the selection on the basis of a desired binding or catalytic activity for the selection of RNA aptamers or ribozymes, respectively. Then, the DNA pool for the next round is re-generated via reverse transcription-PCR by using the selected RNAs as templates. Because the promoter region is eliminated in the RNA pool and recovered by the PCR step, canonical *in vitro* selection procedures cannot be employed directly for the selection of promoter sequences. Breaker and Joyce have developed an *in vitro* selection system for promoter sequences, which is based on coupled transcription/reverse transcription of hairpin-shaped template DNA (Figure 5) [51].

In a previous report, the authors examined the continuous evolution of a ribozyme via isothermal amplification of active ribozymes by using coupled reactions of transcription, ribozyme catalysis, and reverse transcription *in vitro* [52]. During the experiment, they came across the enrichment of inactive, parasitic sequence species lacking the ribozyme sequence [52]. The parasitic DNAs were single-stranded and adopted a hairpin structure that contains a double-stranded promoter region for typical transcription initiation. Because the transcribing RNA polymerase passes around the hairpin region, not only the template strand region but the non-template regions are also transcribed extendedly.

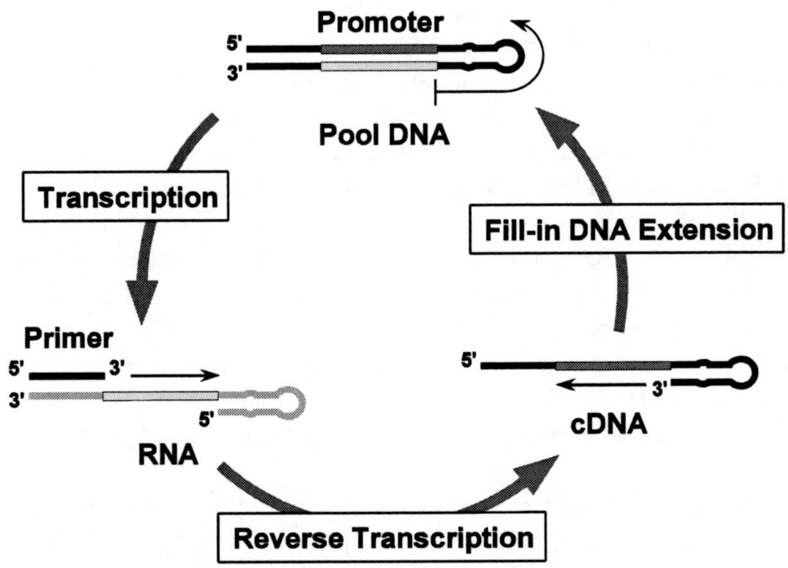

Figure 5. *In vitro* selection of promoter sequences by coupled transcription/reverse transcription of hairpin-shaped template DNAs. The pool DNA is composed of a primer-binding site, promoter sequence to be tested, and terminal loop structure. The transcription generates RNAs containing the template strand sequence of the promoter, and the reverse transcription generates cDNA including the non-template strand of the promoter. The subsequent fill-in DNA extension re-generates the initial sequence of the pool DNAs. Thin arrows indicate the orientations of RNA and DNA syntheses by the RNA polymerase and reverse transcriptase. The sequence regions of template and non-template strands of the promoter are indicated with light and dark gray boxes, respectively.

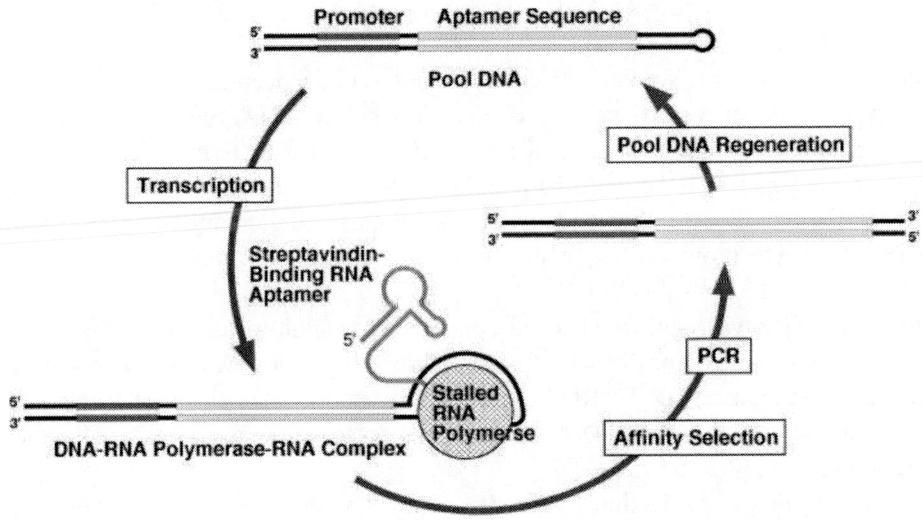

Figure 6. *In vitro* selection of promoter sequences employing hairpin-shaped template DNAs encoding an RNA aptamer. The pool DNA is composed of primer-binding sites, promoter sequence to be tested, the coding sequence of a streptavidin-binding aptamer, and terminal loop structure. The transcribing RNA polymerase is stalled at the loop region, resulting in the formation of DNA-RNA polymerase-RNA ternary complex. Because of the affinity of the RNA aptamer, the complex is purified with streptavidin resin from the mixture with the uncomplexed DNAs. The DNA in the recovered complex is amplified by PCR, and terminal loop structure is added to re-generate the pool DNA for the next round.

Thus, the synthesized RNA contains the template strand sequence of the promoter, and subsequent reverse transcription generates complementary DNA (cDNA) including the non-template strand of the promoter. After dissociation from the RNA, the cDNA is folded into the hairpin structure with a 5'-single-stranded region that provides an intra-molecular template for the DNA-dependent DNA polymerase activity of the reverse transcriptase. Finally, the subsequent fill-in DNA elongation re-generates the parasitic DNA.

By using this parasitic sequence, they examined a process of continuous selection of promoter sequences [51]. Because of the autonomous nature of the amplification process of the parasitic sequence, the system requires only a serial transfer of a part of the reaction mixture to the new mixture every two hours.

The authors examined the selections of promoter sequences for T7, T3, or SP6 RNA polymerases from a library with 17-nt random region, and all three trials with five serial transfers showed successful enrichment of sequences similar to the corresponding wild-type promoters.

Although the Breaker and Joyce's system achieves the promoter sequence selection from very large libraries, and its experimental manipulation is very easy, the system has a restriction on the reaction condition to be tested [53]. Both RNA polymerase and reverse transcriptase need to work in the same buffer and because the reaction uses the enzymes simultaneously, the parasitic sequence cannot be efficiently amplified in modified buffers.

One possible alternative strategy for the *in vitro* selection in bulk solution is the direct recovery of template DNAs that carry the active promoter sequence. The author has developed this strategy for the selection system employing the stalling event of transcribing RNA polymerase on a template DNA [53].

The concatamer junction (CJ) of the T7 DNA acts as a transcription pause site, and at the CJ site, and the transcribing RNA polymerase is temporally stalled at the CJ site without releasing the RNA synthesized [54].

The author found that similar RNA polymerase stalling occurred when a hairpin-shaped DNA with a long stem region was employed as a template DNA. Such stalling events result in the formation of a stable DNA-RNA polymerase-RNA ternary complex; i.e., the synthesized RNA can be tethered to its template DNA via the stalled RNA polymerase [53]. Thus, when the template DNA encodes an RNA aptamer, the ternary complex including the DNA with an active promoter sequence can be recovered using the binding activity of the aptamer.

This possibility was confirmed using aptamer S1, a high-affinity RNA aptamer against streptavidin, whose utility for the affinity purification has been previously demonstrated [55]. The model experiment using T7 RNA polymerase demonstrated that one round of *in vitro* selection enriched the template DNA with T7 promoter sequence over the DNA with SP6 promoter more than 20-fold [53].

In addition, *in vitro* selection of T7 promoter variants was examined from a partially randomized promoter library (six nucleotides at position -11 to -6 of the T7 promoter were randomized). Three rounds of the selection resulted in a promoter activity of the pool DNA, which was comparable with that of the original T7 promoter, and the results of sequence analysis showed that the sequence distribution correlated highly with the published activities of mutated sequences of T7 promoter [53].

Conclusion

Bacteriophage RNA polymerases are promising components for artificial *in vitro* and prokaryotic expression systems. Traditionally, *in vivo* systems have been employed for the analysis and engineering of promoter sequences for bacteriophage RNA polymerases. Because typical *in vivo* systems require common experimental setups and manipulations, they may be more familiar for researchers in various fields of life sciences. In contrast, several *in vitro* systems developed recently enable the examination of huge numbers of promoter sequences. All these systems would contribute to the extension of orthogonal transcription machineries for artificial expressions in synthetic gene circuits.

References

[1] Khalil, A. S. and Collins, J. J. (2010) Synthetic biology: applications come of age. *Nature Rev. Genet.,* 11, 367-379.

[2] Randall, A., Guye, P., Gupta, S., Duportet, X., and Weiss, R. (2011) Design and connection of robust genetic circuits. *Methods Enzymol.*, 497, 159-86.

[3] Nandagopal, N., and Elowitz, M. B. (2011) Synthetic biology: integrated gene circuits. *Science,* 333, 1244-1248.

[4] Chamberlin, M., Kingston, R., Gilman, M., Wiggs, J., and deVera, A. (1983) Isolation of bacterial and bacteriophage RNA polymerases and their use in synthesis of RNA *in vitro*. *Methods Enzymol.*, 101, 5540-68.

[5] Gurevich, V. V. (1996) Use of bacteriophage RNA polymerase in RNA synthesis. *Methods Enzymol.*, 275, 382-397.

[6] Milligan, J. F. and Uhlenbeck, O. C. (1989) Synthesis of small RNAs using T7 RNA polymerase. *Methods Enzymol.*, 180, 51-62.

[7] Krieg, P. A. and Melton, D. A. (1987) *In vitro* RNA synthesis with SP6 RNA polymerase. *Methods Enzymol.*, 155, 397-415.

[8] Tabor, S. and Richardson, C. C. (1985) A bacteriophage T7 RNA polymerase/promoter system for controlled exclusive expression of specific genes. *Proc. Natl. Acad. Sci. USA.,* 82, 1074-1078.

[9] Studier, F. W. and Moffatt, B. A. (1986) Use of bacteriophage T7 RNA polymerase to direct selective high-level expression of cloned genes. *J. Mol. Biol.*, 189, 113-130.

[10] Fuerst, T. R., Niles, E. G., Studier, F. W., and Moss, B. (1986) Eukaryotic transient-expression system based on recombinant vaccinia virus that synthesizes bacteriophage T7 RNA polymerase. *Proc. Nat. Acad. Sci., U. S. A.,* 83, 8122-8126.

[11] Usdin, T. B., Brownstein, M. J., Moss, B., and Isaacs, S. N. (1993) SP6 RNA polymerase containing vaccinia virus for rapid expression of cloned genes in tissue culture. *Biotechniques,* 14, 222-224.

[12] Sagawa, H., Ohshima, A., and Kato, I. (1996) A tightly regulated expression system in Escherichia coli with SP6 RNA polymerase. *Gene,* 168, 37-41.

[13] Studier, F. W. (1991) Use of bacteriophage T7 lysozyme to improve an inducible T7 expression system. *J. Mol. Biol.*, 219, 37-44.

[14] Chelliserrykattil, J., Cai, G., and Ellington, A. D. (2001) A combined in vitro/in vivo selection for polymerases with novel promoter specificities. *BMC Biotechnol.*, 1, 13.

[15] Temme, K., Hill, R., Segall-Shapiro, T. H., Moser, F., and Voigt, C. A. (2012) Modular control of multiple pathways using engineered orthogonal T7 polymerases. *Nucleic Acids Res.*, 40, 8773–8781.

[16] Cheetham, G. M., Jeruzalmi, D., and Steitz, T. A. (1999) Structural basis for initiation of transcription from an RNA polymerase-promoter complex. *Nature*, 399, 80-83.

[17] Cheetham, G. M. and Steitz, T. A. (1999) Structure of a transcribing T7 RNA polymerase initiation complex. *Science*, 286, 2305-2309.

[18] Bertrand, K. P., Postle, K., Wray L. V. Jr., and Reznikoff W. S. (1984) Construction of a single-copy promoter vector and its use in analysis of regulation of the transposon Tn10 tetracycline resistance determinant. *J. Bacteriol.,* 158, 910-919.

[19] Gourse, R. L., de Boer, H. A., and Nomura, M. (1986) DNA determinants of rRNA synthesis in E. coli: Growth rate dependent regulation, feedback inhibition, upstream activation, antitermination. *Cell*, 44, 197-205.

[20] Simons, R.W., Houman, F., and Kleckner, N. (1987) Improved single and multicopy he-based cloning vectors for protein and operon fusions. *Gene*, 53, 85-96.

[21] Lodge, J., Williams, R., Bell, A., Chan, B., and Busby, S. (1990) Comparison of promoter activities in *Escherichia colt and Pseudomonas aeruginosa:* use of a new broad-host-range promoter-probe plasmid. *FEMS Microbiol. Lett.*, 67, 221-226.

[22] Lodge, J., Fear, J., Busby, B., Gunasekaran, P., Kamini, N. R. (1992) Broad host range plasmids carrying the *Escherichia coil* lactose and galactose operons. *FEMS Microbiol. Lett.*, 95, 271-276.

[23] Jensen, P. R. and Hammer, K. (1998) The sequence of spacers between the consensus sequences modulates the strength of prokaryotic promoters. *Appl. Environ. Microbiol.*, 64, 82-87.

[24] Ikeda, R. A., Chang, L. L., and Warshamanat, G. S. (1993) Selection and characterization of a mutant T7 RNA polymerase that recognizes an expanded range of T7 promoter-like sequences. *Biochemistry*, 32, 9115-9124.

[25] Khlebnikov, A., Risa, Ø., Skaug, T., Carrier, T. A., and Keasling, J. D. (2000) Regulatable arabinose-inducible gene expression system with consistent in all cells of a culture control. *J. Bacteriol.,* 182, 7029-7036.

[26] Alper, H., Fischer, C., Nevoigt, E., and Stephanopoulos, G. (2005) Tuning genetic control through promoter engineering. *Proc. Natl. Acad. Sci. U. S. A.,* 12678-12683.

[27] Hicks, B. W. (2002) Green Fluorescent Protein: Applications and Protocols (1st edition; Methods in Molecular Biology). New York, USA; Humana Press.

[28] Shin, I., Kim, J., Cantor, C. R., and Kang, C. (2000) Effects of saturation mutagenesis of the phage SP6 promoter on transcription activity presented by activity logos. *Proc. Natl. Acad. Sci. USA.,* 97, 3890-3895.

[29] Shin, I. and Kang, C. (2003) Mutational analysis and structure of the phage SP6 promoter. *Methods Enzymol.*, 370, 658-668.

[30] Masukata, H. and Tomizawa, J. (1986) Control of primer formation for ColE1 plasmid replication: conformational change of the primer transcript. *Cell,* 44, 125-136.

[31] Kwon, Y. S., Kim, J., and Kang, C. (1998) Viability of *E. coli* cells containing phage RNA polymerase and promoter: interference of plasmid replication by transcription. *Genet. Anal. Biomol. Eng.*, 14, 133-139.

[32] Brendel, V. and Perelson, A. S. (1993) Quantitative model of ColE1 plasmid copy number control. *J. Mol. Biol.*, 229, 860-872.

[33] Esvelt, K. M., Carlson, J. C., and Liu, D. R. (2011) A system for the continuous directed evolution of biomolecules. *Nature,* 472, 499-503.

[34] Riechmann, L. and Holliger, P. (1997) The C-terminal domain of TolA is the coreceptor for filamentous phage infection of *E. coli. Cell*, 90, 351–360.

[35] Calendar, R. (2006) The Bacteriophages (2nd edition). Oxford, UK; Oxford University Press.

[36] Griffiths, A. D. and Tawfik, D. S. (2006) Miniaturising the laboratory in emulsion droplets. *Trends Biotechnol.*, 24, 395-402.

[37] Lu, W. C. and Ellington, A. D. (2012) *In vitro* selection of proteins via emulsion compartments. *Methods*. (In press)

[38] Tawfik, D. S. and Griffiths, A. D. (1998) Man-made cell-like compartments for molecular evolution. *Nature Biotechnol.*, 16, 652-656.

[39] Davidson, E. A., Blarcom, T. V., Levy, M., and Ellington, A. D. (2010) *Pac. Symp. Biocomput.*, 15, 433-443. Emulsion based selection of T7 promoters of varying activity.

[40] Kojima, T., Ohuchi, S., Ito, Y., and Nakano, H. (2012) High-throughput screening method for promoter activity using bead display and a ligase ribozyme. *J. Biosci. Bioeng.*, 114, 671-676.

[41] Dressman, D., Yan, H., Traverso, G., Kinzler, K. W., and Vogelstein, B. (2003) Transforming single DNA molecules into fluorescent magnetic particles for detection and enumeration of genetic variations. *Proc. Natl. Acad. Sci. USA.*, 100, 8817-8822.

[42] Kojima, T., Takei, Y., Ohtsuka, M., Kawarasaki, Y., Yamane, T., and Nakano, H. (2005) PCR amplification from single DNA molecules on magnetic beads in emulsion: application for high-throughput screening of transcription factor targets. *Nucleic Acids Res.*, 33, e150.

[43] Williams, R., Peisajovich, S. G., Miller, O. J., Magdassi, S., Tawfik, D. S., and Griffiths, A. D. (2006) Amplification of complex gene libraries by emulsion PCR. *Nature Methods*, 3, 545–550.

[44] Ikawa, Y., Tsuda, K., Matsumura, S., and Inoue, T. (2004) *De novo* synthesis and development of an RNA enzyme. *Proc. Natl. Acad. Sci. USA.,* 101, 13750-13755.

[45] Ikawa, Y., Matsumoto, J., Horie, S., and Inoue, T. (2005) Redesign of an artificial ligase ribozyme based on the analysis of its structural elements. *RNA Biol.*, 2, 137-142.

[46] Ohuchi, S. P., Ikawa, Y., and Nakamura, Y. (2008) Selection of a novel class of RNA-RNA interaction motifs based on the ligase ribozyme with defined modular architecture. *Nucleic Acids Res.*, 36, 3600-3607.

[47] Paul, S., Stang, A., Lennartz, K., Tenbusch, M., and Uberla, K. (2012) Selection of a T7 promoter mutant with enhanced in vitro activity by a novel multi-copy bead display approach for in vitro evolution. *Nucleic Acids Res.*, (In press)

[48] Stein, V., Sielaff, I., Johnsson, K., and Hollfelder, F. (2007) A covalent chemical genotype-phenotype linkage for *in vitro* protein evolution. *Chembiochem*, 8, 2191–2194.

[49] Ellington, A. D., Chen, X., Robertson, M., and Syrett, A. (2009) Evolutionary origins and directed evolution of RNA. *Int. J. Biochem. Cell Biol.*, 41, 254-265.

[50] Joyce, G. F. (2007) Forty years of *in vitro* evolution. *Angew. Chem. Int. Ed. Engl.*, 46, 6420-6436.

[51] Breaker, R. R., Banerji, A., and Joyce, G. F. (1994) Continuous *in vitro* evolution of bacteriophage RNA polymerase promoters. *Biochemistry*, 33, 11980-11986.

[52] Breaker, R. R. and Joyce, G. F. (1994) Emergence of a replicating species from an *in vitro* RNA evolution reaction. *Proc. Natl. Acad. Sci. U. S. A.*, 91, 6093-6097.

[53] Ohuchi, S. (2012) *In vitro* selection of bacteriophage promoters employing a terminally capped template DNA and a streptavidin-binding aptamer. *J. Biosci. Bioeng.*, 114, 110-112.

[54] Lyakhov, D. L., He, B., Zhang, X. , Studier, F. W., Dunn, J. J., and McAllister, W. T. (1997) Mutant bacteriophage T7 RNA polymerases with altered termination properties. *J. Mol. Biol.*, 269, 28-40.

[55] Srisawat, C. and Engelke, D. R. (2001) Streptavidin aptamers: affinity tags for the study of RNAs and ribonucleoproteins. *RNA*, 7, 632-641.

In: Bacteriophages
Editors: Clark Denton and Richard J. Crosby

ISBN: 978-1-62618-513-5

Chapter 8

The Use of Bacteriophages in Bacterial Food Pathogen Regulation

***Rogeria C. C. Almeida*[1*]*, Lívia P. R. Rossi*[2] *and Paulo F. Almeida*[3]

[1]Universidadc Federal da Bahia, Departamento de Ciência dos Alimentos da Escola de Nutrição, Av. Araújo Pinho, Canela
[2]Instituto Federal de Educação, Ciência e Tecnologia Baiano, Campus Catu. Rua Barão de Camaçari, Barão de Camaçari
[3]Universidade Federal da Bahia, Departamento de Ciências da Biointeração do Instituto de Ciências da Saúde, Av. Reitor Miguel Calmon, Canela

Abstract

Contemporary technologies that are employed to inactivate bacterial pathogens in foods are not flawless; thus, novel approaches for improving food safety are needed. Moreover, consumers desire more natural foods, i.e., free from chemical preservatives. Bacteriophages, which are viruses that infect and kill bacteria, have garnered recent interest for use in food safety. Bacteriophages are components of the natural microflora that are present throughout food production, from the farm to the retail outlet. They are stable in these environments and are readily recovered from soil, sewage, water, farm and processing plant effluents, feces, and retail foods. Studies that have assessed the use of bacteriophages to control food pathogens have yielded promising results but require further improvement. The success of this methodology is ultimately based on consumer acceptance; thus, careful consideration of phage origin must be performed before introducing bacteriophages into foods.

[*] Corresponding author: Tel/Fax: +55 7132837737. E-mail address: rogeriac@ufba.br (R.C.C. Almeida).

Introduction

Bacteriophages are naturally occurring viruses that are suitable candidates for the environmentally friendly biocontrol of pathogens. They have been proposed as antibiotic alternatives for animal health, as biopreservatives in food, and as tools for detecting pathogenic bacteria throughout the food chain.

They are the most abundant self-replicating units in our environment and are present in significant numbers in water and foods of various origins, particularly fermented foods and other products produced with the aid of bacteria. In addition, the phages are typical commensals of humans and animals and are especially abundant in the gastrointestinal tract.

With the advent of antibiotics, phage research was abandoned in most of the Western world. However, the current threat of antibiotic-resistant bacteria has invigorated interest in exploring bacteriophages as biocontrol agents in Western countries. Some studies document the use of phages for controlling undesirable bacteria, which reveal a gap in their use for treating illnesses in human beings and animals.

The failure of Western scientists to achieve consistent results may be attributable to a general lack of understanding of phage biology and replication dynamics. Phage replication critically depends on bacterial density, which acts as a distinct threshold that dictates the increase or decrease in phage number, termed the phage proliferation threshold. The success of phage therapy also depends on various life history parameters, including the inoculum size and timing.

The potential for phage-mediated control of foodborne pathogens has been documented in recent studies with *Salmonella* spp., *Campylobacter* spp., *Escherichia coli*, and *Listeria monocytogenes*.

Phages as Antimicrobial Agents

Bacteriophages—Greek for "bacteria eaters"—were discovered almost a century ago. In 1915, Frederick W. Twort published a note in England that described the infectious destruction of bacterial cultures, producing small cleared areas for filterable entities. They were inactivated by heating at 60°C for 1 hour and could not proliferate independently. In 1917, Felix d´Herelle, at the Pasteur Institute in Paris, independently studied the phenomenon, demonstrating the nature of the particle, which he termed the bacteriophage following the discovery of lytic phages attacking the dysentery bacillus (*Shigella*) in the feces of convalescing patients. From these comments, the phenomenon was termed the Twort-d' Herelle phenomenon.

The viral characteristics of bacteriophages were confirmed by the work of Burnet and Schlsinger in the 1930s. Bacteriophage studies continued through 1943, during which time Felix and Callow isolated the virulent bacteriophage denominated Felix 01.

Bacteriophages do not harbor their own metabolism and rely on host bacteria for multiplication. They exhibit narrow host ranges and typically target specific strains; moreover, they generally do not cross taxonomic boundaries. The first step in the phage reproduction cycle is the recognition of the host-specific primary receptor, which is followed by adsorption to the secondary receptor. The first attachment is reversible because the target cell may not be alive, precluding permanent attachment. The second attachment step is irreversible. Subsequently, the phage DNA circularizes, and the bacterium initiates phage protein production. These proteins sequester the entire host cell to exclusively produce new phages. After progeny phage assembly, two separate phage proteins degrade the cell envelope (lysis), thereby releasing daughter phages that are poised to initiate the next cycle.

The requirement for phage/host system compatibility for each step in the infection process yields an extremely high specificity of phages for distinct hosts. For example, the phage must be able to recognize a cell wall as useless if the phage genome cannot be interpreted by that host. Additionally, every bacterium harbors its own complement of specialized phages that compete with foreign phages, which could render non-specialized intruder phages ineffective. In the absence of target bacteria, the phages decompose into common biological particles that are naturally reabsorbed into the environment.

The inherent phage specificity results in the exclusive elimination of the target organisms without compromising the viability of other, autochthonic bacteria in the habitat. Notably, a successful phage infection inevitably leads to cell death. Therefore, if contaminating bacteria in food encounter a bacteriophage and become infected, the bacteria will certainly die. In ecological terms, phage populations can only increase when a sufficient host amount is present. However, bacteriophages will be differently employed in food safety, where the goal is to eradicate pathogens to prevent further pathogen outgrowth.

Like all viruses, phages are composed of either DNA- or RNA-based genomes, which have been observed in both single- and double-stranded forms. The structure of bacteriophages is similar to other living organisms: a polynucleotide chain consisting of a deoxyribose (or ribose) phosphate backbone attached to the four nucleotides - adenine, thymine (or uracil), guanine and cytosine. With the exception of single-stranded phages, all phages display two complementary polynucleotide chains that are paired together in a double helix. The DNA or RNA is covered with a protein layer called a capsid, which is covered, in some phages, by a tail sheath (*tsh*) composed of lipids that likely mediates the adherence of the phage to the host cell wall. The strongest phage promoters, ones that yield the highest levels of protein synthesis, are often those that drive late-expressing genes that encode major structural proteins of the virion. The identity of their locations, sequences, and regulatory elements is useful not only for understanding the basic biological processes that govern the phage-host interaction but also for practical applications, such as efficiently and tightly regulating the expression of heterologous genes in the host.

Phages can undergo two developmental cycles within the host bacteria—lytic or lysogenic. The lytic cycle is composed of a series of events that occur between the attachment of the phage particle to a bacterial cell and the subsequent release of daughter phage particles. Conversely, the lysogenic cycle includes the replication of phage and host genomes for several generations without major metabolic consequences for the cell. This is a latent and rare mode of infection. In this state, the phage genes occasionally revert to the lytic cycle, resulting in a release of phage particles. This property is known as lysogeny, and phages that can develop both lytically and lysogenically are termed temperate phages.

Bacteriophages that harbor double-stranded DNA (dsDNA) comprise the most abundant group of phages in the biosphere, and approximately 4,500 dsDNA phages can infect a great diversity of bacterial hosts. The classification of bacteriophages based on their host group, morphology, or life cycles has led to conflicting conclusions regarding their origins and evolution.

Recent advances in genomics and phylogeny have elucidated gene flow mechanisms among phages and hosts, which permit researchers to avoid or re-design potentially harmful bacteriophages without undesirable traits or gene dissemination systems. For example, bacteriophages could be genetically engineered to block phage replication once the host has been killed, which would prevent the release of large numbers of phages.

Strategies for Controlling Food Pathogens Using Bacteriophages

Over the last few years, several strategies to minimize the microbial load of raw products have been explored given that the use of antibiotics is restricted because of their negative impact on human antimicrobial therapies. Problems with organoleptic acceptability and its deterioration have been described after physical treatments such as steam, dry heat, and UV light. Moreover, the extensive use of sanitizers has yielded resistant bacteria, which has rendered these procedures less effective. Conversely, common approaches used in the food industry to reduce contamination by foodborne pathogens cannot be directly applied to fresh fruits, vegetables and ready-to-eat products. Recent phage molecular biology research has fuelled multiple biotechnological applications in diverse fields, including nanotechnology, vaccine development, therapeutic delivery, bacterial detection systems, and the development of novel antimicrobials to combat antibiotic-resistant bacteria. The use of phages as natural antimicrobials in food to inhibit undesirable bacteria has also shown promise, which is likely to be accepted by consumers.

In addition to phage therapy, water and food safety, agriculture and animal health are potential application targets that may benefit from phage interventions. Several new companies are considering the use of bacteriophages as antimicrobial agents and tools for detecting pathogens in feed and foodstuffs. OmniLytics, Inc. gained US Environmental Protection Agency approval in 2006 to use its product AgriPhage to treat tomato and pepper spots. In 2007, OmniLytics, Inc. received FDA approval for an anti-*E. coli* hide wash and an anti-*Salmonella* wash/mist/spray to treat live animals prior to slaughter. The USFDA recently approved the use of a bacteriophage prepared from six individually purified bacteriophages as an antimicrobial agent against *L. monocytogenes* in ready-to-eat meat and poultry products. The approved bacteriophage preparation has been shown to be effective against 170 strains of *L. monocytogenes*. The commercial product LISTEX_P100 was approved as a food biopreservative and was granted GRAS (Generally Recognized as Safe) status. Additionally, Listex has been approved for use in Switzerland in cheese-making and as processing aids to satisfy European food safety legislation.

Bacteriophage-based biocontrol measurements exhibit a great potential to enhance microbiological safety based on their long history of safe use, relatively easy handling and high and specific antimicrobial activity.

Using phages to combat food pathogens can be employed at all stages of production, including the classic 'farm to fork' approach and throughout the entire food chain. The farm to fork concept identifies quality assurance steps where bacterial contamination may occur and where phage treatments may be applied. The most frequently encountered food pathogens belong to one of the four dominant genera, *Salmonella*, enterotoxigenic *Escherichia coli*,

Campylobacter, and *Listeria*, and less common infections by *Clostridium* spp., *Staphylococcus aureus*, *Streptococcus suis*, and *Cronobacter sakazakii* have been documented.

Bacteriophages are suitable for (i) preventing or reducing colonization and disease in livestock (phage therapy); (ii) decontaminating carcasses and other raw products, such as fresh fruit and vegetables, and for disinfecting equipment and contact surfaces (phage bio-sanitation and biocontrol); and (iii) extending the shelf life of perishable manufactured foods such as natural preservatives.

The broad-host-range bacteriophages A511 and P100 have been previously used to control *L. monocytogenes* in foods. In liquid foods, such as chocolate milk and mozzarella cheese brine, bacterial counts rapidly dropped below the level of direct detection. In solid foods, bacteriophages reduced bacterial counts by up to five log units. The authors also considered applying more bacteriophage particles (3×10^8 pfu/g) to attain more effective results than lower doses did.

Oral administration of a high dose of phage P100 in male and female Wistar albino rats revealed no in-life effects attributable to the material. None of the animals died prematurely or displayed abnormal physical signs or behavioral changes throughout the entire observation period.

Studies assessing the safety of phage application in animals also reported no adverse or unexpected effects. Consistently, a recent study observing human volunteers receiving phage T4 concluded that it was safe for oral administration; phage or phage-specific antibodies could not be detected in the serum of the human subjects. In conclusion, evidence suggests that the intake of phage with food does not exhibit any negative effects on humans.

Despite these great advantages, some authors have proposed arguments against using of bacteriophages as antimicrobial agents. The mechanism used to transduce phages can transfer undesirable characteristics, such as virulent genes, from one organism to another, and the lysogenic conversion can produce bacterial cells that are less susceptible to phages. Therefore, before phages are employed for biocontrol, their transduction mechanisms must be carefully assessed to determine whether they can transfer virulence genes that could be expressed in the host before cell death.

The emergence of phage-resistant bacteria is one of the primary concerns in food system management, which has been noted for foodborne pathogens in the pre-slaughter environment and for spoilage bacteria during food storage. However, phages are constantly evolving and may have the potential to counteract this resistance. Using a phage mixture from distinct families may also help combat bacterial resistance.

Little is known about the emergence frequency of these phage-resistant bugs, suggesting the need for more extensive analysis when selecting phages for potential use. However, although this is a pertinent issue in the acceptability of phages for human consumption, phage-resistant mutants in *Salmonella* Enteritidis strains have been shown to become avirulent through the loss of the O-Polysaccharide layer, which is required for phage adsorption. A similar finding has also been observed in the fish pathogen *Pseudomonas pecoglossida*. However, resistant remains a concern and requires evaluation on an individual basis.

To guarantee proper phage performance, phages should be isolated from the same niche / habitat to maximize the likelihood of the phage to replicate and survive in those conditions.

Phage antimicrobial activity that has been observed in laboratory conditions could be greatly reduced in food systems. Factors that limit this reduction include decreased diffusion rates, which diminish the chance of host-phage collisions, the microbial load, which may act as a mechanical barrier by providing unspecific phage binding sites, and other adverse factors such as temperature, pH and inhibitory compounds.

There are several strategies that can improve and expand the antimicrobial activity of phages. Comparative genomics of phage tail fiber genes involved in recognizing specific host receptors may lead to approaches that expand the host range. A detailed analysis of bacterial receptors may also help to elucidate and predict the development of bacteria-insensitive mutants, which often occur due to the loss or mutation of these receptors. Both approaches may enhance the use of phages as antimicrobials and as bacterial detection systems. New phage-derived antibacterial strategies based on phage-encoded proteins that drive cellular-mediated phage proliferation, such as endolysins, are currently being evaluated.

Bacteriophage Survival in Foods and on Facility Surfaces

Phage adsorption to receptors on bacterial cell walls depends on a stochastic collision between the phage and a bacterial cell. This initial interaction can be impeded by several physical barriers. For example, the viscous environment of the intestine or rumen could protect bacteria from attack, and the particulate nature of many food environments or non-target indigenous bacterial contaminants readily form physical barriers that interfere with phage-bacteria interactions.

Treatment of facility surfaces is also a plausible application for phages. Food-contact surfaces, in particular, could be cleaned effectively using phages, even during production, without interrupting manufacturing processes. Non-food-contact surfaces could also be treated using phages; however, phages cannot reach certain places that aggressive chemicals can.

Other ecological considerations in phage biocontrol strategies include the pH of the food or animal host environments, temperature, immune responsiveness, and the *in vivo* physiology of the bacterial strains. A thorough understanding of these environmental influences may elucidate the disparity among phage *in vitro* and *in vivo* lytic activities and enable the design of more effective phage biocontrol strategies.

Food matrix: A successful phage infection and subsequent killing of the host cells strongly depends on the environmental conditions, i.e., the type of food and its specific matrix. A high density of phage particles increases the probability of non-motile phage particles to reach the target cells given that they do so exclusively by diffusion. More specifically, the optimum phage concentration should be at least 10^8 PFU/g or cm^2 at the time of application. This phage concentration is sufficient for contact between the passively diffusing virus particles and their host cells within the given time and spatial limitations. In liquid foods, this does not appear to be a problem because suspended phage particles can diffuse almost freely. Conversely, the total surface area and its ability to absorb liquid from the phage suspension are the decisive parameters in solid foods with an even surface.

Moreover, target bacteria may be embedded within the complex food matrices, thereby shielding them from diffusing phage particles.

pH tolerance: Bacteriophages are generally stable in pH values between 5.0 and 8.0 and between 4.0 and 10.4 at low temperatures. An assessment of *Salmonella* in phage-treated fresh fruits indicated that the phage stability was pH-dependent. When phages were incubated at 5°C for 24 hours at a pH of 4.2, the phage number was four times lower than in preparations with a pH of 5.8. Mixing phages with *Listeria monocytogenes* controlled the number of bacteria in fresh cuts of cantaloupes. Conversely, *Salmonella* Enteritidis and *L. monocytogenes* growth on artificially inoculated apple slices was unaffected by phage treatment. Because the pH was not in a critical range, the phage particles may have been inactivated due to secondary plant compounds or substances known to inactivate viruses and bacteriophages, such as organic acids and tannins.

Thermal tolerance: The thermal inactivation of phages in laboratory conditions generally occurs between 60 and 75°C, but thermal stress can exhibit variable effects on the phages. Bacteriophages remain problematic in fermented dairy production, where a process to eliminate contaminants completely from the fermentation environment is lacking. Processes such as pasteurization do not adequately eliminate bacteriophages. The degree of thermal resistance depends on the phage considered and the culture media used. Phages isolated from the serum of cheese and yogurt required heating at 90°C for 5 minutes to reduce phage numbers from 10^6 - 10^7 PFU/mL to less than 10 PFU/mL. Phages can remain viable for extended periods of time in refrigerated foods and can subsequently initiate the infectious cycle when susceptible host bacteria grow to sufficient numbers.

Visible and UV light: Current evidence suggests that the majority of phages become exponentially inactivated after exposure to ultraviolet light. An accumulation of nucleic acid damage likely causes the majority of the inactivation. Sunlight appears to be the most important mechanism for inactivating sewage microorganisms in shallow seawater, although different rates of inactivation for various bacteriophages have been reported. In aquatic systems, the solar light (particularly UV-B) may be the primary factor governing phage number decline. Phages irradiated with UV have been shown to reactivate after repairing their nucleic acid via triggering a light-dependent host cell repair mechanism or, given multiple infections, via recombination between clones.

Sanitizer: Studies in dairy factories have demonstrated that sodium hypochlorite (100 ppm of free chlorine) and peracetic acid can inactivate phages in milk within 5 to 10 minutes. The use of 75% ethanol (v/v) was also effective against phages, whereas isopropanol was not as efficient. Given that the phages are generally more resistant than bacteria to ambient changes in food processing, they can persist in places that were not adequately cleaned or sanitized.

Conclusion

The success of using phages to combat bacteria depends on the emergence of or persistent resistance to viruses, similar to the emergence of antibiotic resistance. Most research has been conducted with experimentally infected foods; thus, field trials on naturally infected foods are needed. Once conditions for an effective phage biocontrol method have been optimized, demonstrating that phages can reduce the persistence of pathogens in the environment and limit the transfer from animals to foods during the initial stages of processing remains critical. Of all of the factors that can limit the ability of phages to control indigenous bacteria, the restricted diversity of susceptible bacteria in foods is the most pressing concern. A more rewarding and expedient approach may be to breed phages with broader host ranges and to assemble them into a cocktail with complementary activities. The final measure of whether phage applications in foods will be successful is the consumer. Therefore, the sources of all phages that are considered in food application must be extensively evaluated.

References

Ackermann, HW; Greer, GG; Rocourt, J. Morphology of *Brochothrix thermosphacta* phages. *Microbios*, 1988, 56, 19-26.

Atterbury, RJ; Connerton, PL; Dodd, CE; Rees, CE; Connerton, IF. Application of host-specific bacteriophages to the surface of chicken skin leads to a reduction in recovery of *Campylobacter jejuni*. *Appl Environ Microbiol*, 2003, 69, 6302-6.

Avsaroglu, D; Bozog˘lu, F; Akçelik, M; Bayindirli, A. Effect of high pressure on lactococcal bacteriophages. *Journal of Food Safety*, 2009, 29, 26–36.

Berchieri, Jr, A; Lovell, MA; Barrow, PA. The activity in the chicken alimentary tract of bacteriophages lytic for *Salmonella typhimurium*. *Res. Microbiol*, 1991, 142, 541–549.

Biswas, B; Adhya, S; Washart, P; Paul, B; Trostel, AN; Powell, B; Carlton, R; Merril, CR. Bacteriophage therapy rescues mice bacteremic from a clinical isolate of vancomycin-resistant *Enterococcus faecium*. *Infect Immunol*, 2002, 70, 204–210.

Bruttin, A; Brussow, H. Human volunteers receiving *Escherichia coli* phage T4 orally: a safety test of phage therapy. *Antimicrob Agents Chemother,* 2005, 49, 2874-78.

Carlton, RM; Noordman, WH; Biswas, B; Meester, ED; Loessner, MJ. Bacteriophage P100 for control of *Listeria monocytogenes* in foods: Genome sequence, bioinformatic analyses, oral toxicity study, and application. *Reg Toxic Pharmacol*, 2005, 43, 301-12.

Cerveny, KE; De Paola, A; Duckworth, H; Gulig, PA. Phage therapy of local and systemic disease caused by *Vibrio vulnificus* in irondextran-treated mice. *Infect. Immunol*, 2002, 70, 6251–6262.

Chibani-Chennoufi, S; Bruttin, A; Dillmann, ML; Brussow, H. Phage-host interaction: an ecological perspective. *J Bacteriol*, 2004, 186, 3677-86.

Chibani-Chennoufi, S; Sidoti, J; Bruttin, A; Kutter, E; Sarker, S; Brüssow, H. In vitro and in vivo bacteriolytic activities of *Escherichia coli* phages: implications for phage therapy. *Antimicrob Agents Chemother*, 2004, 48, 2558–2569.

Dykes, GA; Moorhead, SM. Combined antimicrobial effect of nisin and listeriophage against *Listeria monocytogenes* in broth but not in buffer or on raw beef. *Int J Food Microbiol*, 2002, 73, 71-81

Favrin, SJ; Jassim, SA; Griffiths, MW. Development and optimization of a novel immunomagnetic separation-bacteriophage assay for detection of *Salmonella enterica* serovar Enteritidis in broth. *Appl Environ Microbiol*, 2001, 67, 217-224.

Favrin, SJ; Jassim, AS; Griffiths, MW. Application of a novel immunomagnetic separation-bacteriophage assay for the detection of *Salmonella enteritidis* and *Escherichia coli* O157:H7 in food. *Int J Food Microbiol*, 2003, 85, 63-71.

Garcia, P; Martinez, B; Obeso, JM; Rodriguez, A. Bacteriophages and their application in food safety. *Lett App Microbiol*, 2008, 47, 479-85.

Goode, D; Allen, VM; Barrow, PA. Reduction of experimental *Salmonella* and *Campylobacter* contamination of chicken skin by application of lytic bacteriophages. *App. Environ. Microbiol*, 2003, 69, 5032–5036.

Greer, GG. Bacteriophage control of foodborne bacteria. *J Food Prot*, 2005, 68, 1102-11.

Guenter, S; Huwyler, D; Richard, S; Loeesner, MJ. Virulent bacteriophage for efficient biocontrol of *Listeria monocytogenes* in ready-to-eat foods. *Appl Environ Microbiol*, 2009, 75, 93-100.

Hagens, S; Loessner, MJ. Mini-Review. Application of bacteriophages for detection and control of foodborne pathogens. *Appl Microbiol Biotechnol*, 2007, 76, 513-519.

Hagens, S; Offerhaus, ML. Bacteriophages – New weapons for food safety. *Food Technol,* 2008, 62, 46-54.

Hudson, JA; Billington, C; Carey-Smith, G; Greening, G. Bacteriophages as biocontrol agents in food. *J Food Prot*, 2005, 68, 426-37.

Huff, WE; Huff, GR; Rath, NC; Balog, JM; Donoghue, AM. Alternatives to antibiotics: utilization of bacteriophage to treat colibacillosis and prevent foodborne pathogens. *Poultry Sci*, 2005, 84, 655–9.

Joerger, RD. Alternatives to antibiotics: bacteriocins, antimicrobial peptides and bacteriophages. *Poult Sci*, 2003, 82, 640-7.

Joklik, WK; Willente, HP; Amos, DB, Wilfert, CM. Zinsser microbiología. 20ª ed. Bogota: Panamericana, 1994.

Leverentz, B; Conway, WS; Alavidze, Z; Janisiewicz, WJ; Fuchs, Y; Camp, MJ; et al. Examination of bacteriophage as a biocontrol method for *Salmonella* on fresh-cut fruit: a model study. *J Food Prot*, 2001, 64, 1116-21.

Leverentz, B; Conway, WS; Camp, MJ; Janisiewicz, W; Abuladze, T; Yang, M; et al. Biocontrol of *Listeria monocytogenes* on fresh-cut produce by treatment with lytic bacteriophages and a bacteriocin. *Appl Environ Microbiol*, 2003, 69, 4519-26.

Leverentz, B; Conway, WS; Janisiewicz, W; Camp, MJ. Optimizing concentration and timing of a phage spray application to reduce *Listeria monocytogenes* on honeydew melon tissue. *J Food Prot*, 2004, 67, 1682-6.

Loc Carrillo, C; Atterbury, RJ; El-Shibiny, A; Connerton, PL; Dillon, E; Scott, A; et al. Bacteriophage therapy to reduce *Campylobacter jejuni* colonization of broiler chickens. *Appl Environ Microbiol*, 2005, 71, 6554–63.

Loessner, MJ; Scherer, S. Organization and transcriptional analysis of the *Listeria* phage A511 late gene region comprising the major capsid and tail sheath protein genes *cps* and *tsh*. *J Bacteriol*, 1995, 177, 6601-09.

Mahony, J; McAuliffe, Ross, RP; van Sinderen, D. Bacteriophages as biocontrol agents of food pathogens. *Current Opinion in Biotechnology*, 2010, 22, 1–7.

Mathur, MD; Vidhani, S; PL Mehndiratta, PL. Bacteriophage Therapy: An Alternative to Conventional Antibiotics. *JAPI*, 2003, 51, 593-6.

O'Flynn, G; Ross, RP; Fitzgerald, GF; Coffey, A. Evaluation of a cocktail of three bacteriophages for biocontrol of *Escherichia coli* 0157:H7. *Appl Environ Microbiol*, 2004, 70, 3417-21.

Oliveira, IC; Almeida, RCC; Hofer, E; Almeida, PF. Bacteriophage amplification assay for detection of *Listeria* spp. using virucidal laser treatment. *Brazilian J Microbiol*, 2012, 43, 1128-36.

Rossi, LPR; Almeida, RCC; Lopes, LS; Figueiredo, ACL; Ramos, MPP; Almeida, PF. Occurrence of *Listeria* spp. in Brazilian fresh sausage and control of *Listeria monocytogenes* using bacteriophage P100. *Food Control*, 2011, 22, 954-958.

Sinton, LW; Hall, CH; Lynch, PA; Davies-Colley, RJ. Sunlight inactivation of fecal indicator bacteria and bacteriophages from water stabilization pond effluent in fresh and saline waters. *Appl Environ Microbiol*, 2002, 68, 1122-31.

Siqueira, RS; Dodd, CE; and Rees, CE. Evaluation of the natural virucidal activity of teas for use in the phage amplification assay. *Int J Food Microbiol*, 2006, 111, 259–262.

Sklar, IB; Joerger, RD. Attempts to utilize bacteriophage to combat Salmonella enterica serovar Enteritidis infection in chickens. *J Food Safety*, 2001, 21, 15-29.

Stewart, GSAB; Jassim, SAA; Denyer, SP; Rees, CED; Park, S; Rostas-Mulligan, K. *Methods for rapid microbial detection*. PCT patent WO 92/02633, 1992.

Stewart, GS; Jassim, SA; Denyer, SP; Newby, P; Linley, K; Dhir, VK. The specific and sensitive detection of bacterial pathogens within 4 h using bacteriophage amplification. *J Appl Microbiol*, 1998, 84, 777–83.

Sulakvelidze, A; Alavidze, Z; Morris, JGJ. Bacteriophage therapy. *Antimicrob Agents Chemother*, 2001, 45(3), 649-59.

Sulakvelidze, A; Barrow, P. Phage therapy in animals and agribusiness. In: Kutter, E, Sulakvelidze A. *Bacteriophages: biology and applications*. Boca Raton: CRC Press; 2005, p.335-80.

Tanji, Y; Shimada, T; Yoichi, M; Miyanaga, K; Hori, K; Unno, H. Toward rational control of *Escherichia coli* 0157:H7 by a phage cocktail. *Appl Microbiol Biotechnol.*, 2004, 64, 270-4.

Toro, H; Price, SB; McKee, S; Hoerr, FJ; Krehling, J; Perdue, M; et al. Use of bacteriophages in combination with competitive exclusion to reduce *Salmonella* from infected chicken. *Avian Dis*, 2005, 49, 118-24.

Whichard, JM; Sriranganathan, N; Pierson, FW. Suppression of *Salmonella* growth by wild-type and large-plaque variants of bacteriophage Felix O1 in liquid culture and on chicken frankfurters. *J Food Prot*, 2003, 66, 220-5.

In: Bacteriophages
Editors: Clark Denton and Richard J. Crosby
ISBN: 978-1-62618-513-5

Chapter 9

The Interaction between Bacteriophage λ and Its Bacterial Host Receptor, LamB

***Sujoy Chatterjee*[*] *and Eli Rothenberg*[†]**
Department of Biochemistry and Molecular Pharmacology,
NYU Medical School, New York, NY, US

Abstract

With the recent emergence of antibiotic resistant bacteria, there is an urgent need for new therapeutic alternatives. Bacteriophages are bacterial viruses that provide an attractive alternative to treat bacterial infections. One of the key factors for a successful phage therapy is the high affinity binding of phages onto the bacterial cell surface. This very first step of virus-host interaction would determine subsequent infection steps and the fate of the entire infection process. Phage λ and its host *Escherichia coli* serve as an excellent paradigm to study the mechanism of phage adsorption onto the host cell surface receptor. Here we review our current understanding of the mechanism of lambda phage adsorption with an emphasis on recent imaging studies using fluorescently labeled phage λ and its host receptor LamB.

Introduction

Ever since Alexander Fleming discovered Penicillin, antibiotic therapy has been the main stay against infectious diseases and for sometime it appeared that the battle against infectious disease has been won. But this idea is seriously jeopardized by the recent emergence and

[*] Sujoy Chatterjee. Department of Biochemistry and Molecular Pharmacology, NYU Medical School, 550 First Avenue, New York, NY 10016, US. E-Mail: sujoy.chatterjee@nyumc. org.
[†] Eli Rothenberg. Department of Biochemistry and Molecular Pharmacology, NYU Medical School, 550 First Avenue, New York, NY 10016, US. E-mail: Eli.Rothenberg@nyumc.org.

spread of antibiotic resistance. In the recent past we have witnessed the global dissemination of multi-drug resistant virulent MRSA (methicillin-resistant *Staphylococcus aureus), Acinetobacter baumannii, Klebsiella pneumoniae, Enterobacteriacea, Mycobacterium tuberculosis etc.* [1]. We have come almost full circle and arrived at a point as frightening as pre-antibiotic era. The worldwide spread of pathogenic bacteria that are resistant to a variety of antibiotics threatens to reduce modern medicine to a state reminiscent of the pre-antibiotic era. Even though novel antibiotics directed against such drug resistant bacteria can be developed, the pathogens ultimately become resistant to such drugs. To break this vicious cycle, it will be necessary to adopt chemotherapy-independent remedial strategies to combat bacterial infections.

Bacteriophages (phages) are viruses that specifically infect and lyse bacteria. Phage therapy is a method where phages are used to treat bacterial infections. It was first introduced by Félix d'Herelle, in about 1920. This seminal discovery occurred about 20 years before the discovery of penicillin.

At inception, phage therapy was considered as a promising tool against bacterial infection and was successfully used to combat bacterial infection of wounded soldiers during war in the former Soviet Union and Eastern Europe, but was ultimately abandoned by the West in the 1940s with the advent of the antibiotic era [2]. However, the emergence of multi drug resistance pathogenic bacteria has recently motivated scientific community to reevaluate the alternate approach of phage therapy [3].

Phage therapy has the following advantages over chemotherapy:

1 Selectivity (phages are highly selective for their host bacteria).
2 Minimum cross reactivity (bacteriophages do not affect eukaryotic cells; side effects from phages are minimal).
3 Effectivity (effective against multidrug-resistant pathogens and also able to mutate itself in response to the appearance of phage-resistant mutants).
4 Economic (the cost of developing a phage system is cheaper than developing a new antibiotic).

To develop successful phage therapy, an understanding of the host-virus interaction is a pre-requisite. The life cycle of a lytic bacteriophage involves the following steps:

1. Adsorption of the phage on the bacterial cell through binding to specific receptor onto the host cell surface followed by injection of its nucleic acid inside the host cell.
2. Expression of the early phage genes, which are necessary for the shutting down of the host machinery and phage genome integration.
3. Replication of the phage genome concurrent with the host.
4. Expression of the late genes required for the formation of new phage particles and lysing the host.
5. Assembly of the phage heads and tails and packaging of the phage genome.
6. Release of the new phage progeny by lysing the host. Adsorption is first and foremost step in virus recognition of the host cell [4-6]. The specificity of phage infection is defined at this moment and ultimately determines the fate of infection. Here we review the studies of this interaction using *Escherichia coli* bacterium and

its virus, bacteriophage λ as a virus-host model system with a special emphasis on recent imaging studies developed to track virus target searching mechanisms.

Bacteriophage Receptors on Cell Surface - the Gateway of Phage Entry

The adsorption of phages onto bacterial surface is the first step in the infection process. At this stage, phages have two choices, either dissociate from the host cell, known as desorption or alternatively it binds irreversibly to the host cell, triggers a series of events and finally injects its nucleic acid into the host cytoplasm. The receptors present onto the cell surface are one of the determinants of efficacy of this process. Below is a brief review of phage receptors present in diverse taxonomic groups.

The cell envelope of gram-negative bacteria has two membranes; the outer membrane (OM) and the cytoplasmic membrane (CM) and a thin peptidoglycan layer between them [7, 8]. The outer membranes of gram negative bacteria protect them from harmful agents by slowing down their penetration while allowing uptake of nutrients. The asymmetric outer membrane of gram-negative bacteria differs in structure from their inner membrane; composed of an inner leaflet of mostly phospholipids and the outer lipopolysaccharides (LPS). One of the distinctive features of the OM is its high permeability due to the presence of transport channels formed by the integral proteins. Proteins localized in membrane and various LPS sites serve as receptor for bacteriophage. The major proteins of outer membrane are β-barrel proteins, which play roles both in adhesion and virulence and contribute ~2%–3% of their proteome [9]. However, their occurrence is not limited to gram negative bacteria and also found in mycobacteria, mitochondria chloroplasts [10-13], in several microbial toxins such as heptameric pore-forming α-haemolysin from *Staphylococcus aureus*, toxin aerolysin and the anthrax-protective antigen [14-16].

Porins are the most well characterized outer membrane proteins of gram-negative bacterium. They form general diffusion pore of defined exclusion size for hydrophilic solutes within the outer membrane. In addition to the constitutive porins, the outer membranes also contain inducible porins which are produced only under special growth conditions. The most abundant porins in *Escherichia coli*, OmpF and OmpC porins, are called general porins. OmpC serves as a receptor for phages Tulb, Hy2, ss4 [17, 18]. However phage T4 utilizes both OmpC and cell wall LPS as receptor, mutation in any one of these receptors results in reduced efficiency of infection and loss of both receptors results complete resistance. OmpF is the receptor for phage T2 [19, 20]. Outer membrane proteases can also serve as receptor, e.g. OmpT and OmpX are receptors for T-like phages [21, 22]. In addition to proteins, LPS of the outer membrane in gram-negative bacteria can also act as a receptor for bacteriophage adsorption. LPS is a complex polymer composed of monosaccharides and fatty acids and can be structurally divided into three parts: a lipid moiety known as the lipid A, a non repeating "core" oligosaccharide, and the outermost part – a hydrophilic polysaccharide called the O-antigen which differs between species [23, 24]. LPS can be classified in two groups based on the presence of O-antigen: Smooth (S) type is characterized by typical LPS structure, i.e. comprising lipid A, core and side chain and Rough (R) type lacks O-chain but contains lipid A and the core. Phages specific to S-type LPS display extremely narrow host range due to the diversity of O-antigen structure in bacteria of different taxonomic groups. On the otherhand, bacteriophages use R-type show a broader host range since the structure of LPS core is relatively conserved in gram-negative bacteria.

In contrary, gram-positive bacteria do not possess an outer membrane and their cell wall differs greatly in structure and chemical composition, highly crosslinked and appears as three dimensional mesh-like structures, from the gram-negative bacteria. Peptidoglycan is the main component of their cell wall, which is a heteropolymer composed of disaccharide monomer formed by N-acetylglucosamine and N-acetylmuramic acid. Gram-positive cells also have cell wall glycopolymers (CWGs) which are either attached to the peptidoglycan or to the cytoplasmic membrane [25]. Teichoic acids, teichuronic acids, cell wall polysaccharides and arabinogalactans are examples of peptidoglycan anchored CWGs, while membrane-anchored CWGs are lipoteichoic acids, lipomannans, and lipoarabinomannans. Phages can use peptidoglycan, teichoic/lipoteichoic acids or proteins of the cytoplasmic membrane as their receptors. Phages like sk1, jj50 and A511 use the peptidoglycan; ϕ25 and ϕ29 use teichoic acid components, while lactobacillus phages LL-H and JCL1032 bind to the lipoteichoic acids [26-28]. Bacterial pilli and flagella can also serve as phage receptor. For example, filamentous ssDNA phages M13 and ssRNA phage Qβ use bacterial pili and bacteriophages SP3, PBS1, χ use flagella as their receptors [29-31].

Bacteriophage λ and Its *E. Coli* Receptor LamB

Since its discovery, bacteriophage λ, has served as an excellent paradigm for studying fundamental principles of biology such as gene regulation, DNA replication, homologous and site-specific recombination [32, 33]. The best-characterized feature of phage λ biology is the lytic-lysogenic switch which decides whether the phage propagates by host lysis, or integrates into host genomes. The interaction of the λ with the host *E. coli* cell is other well studied aspect of phage λ biology. In the last century, wealth number of information has been obtained about this interaction both from biochemical studies using purified components and genetically from either mutated phage or bacterial receptor [34-45]. However, a quantitative understanding of the viral target-finding process and its dependence on virus-host receptor interactions and cellular architecture was obtained from recent imaging studies [46, 47].

E. coli maltose porin LamB (λ-receptor) serves as the receptor for Phage λ. Lambda phage resistant mutants of the *E. coli* K-12 are located in two genetic regions *malA* and *malB* [38, 39]. The *malB* region contains a gene *lamB*, which synthesizes the protein, LamB. LamB is a component of *E. coli* outer membrane which acts as a receptor for the phage λ. However, mutant bacteriophage λ can use alternative receptor. For example, when phage cI26 (a lytic derivative of phage λ) was cultured with *E. coli* under conditions where LamB is not expressed, mutant phage use a new receptor, OmpF. A combination of four complimentary mutations in phage tail protein J was required to use this new receptor. At the same time, some host mutations prevented phages from evolving this new function, suggesting complex interactions in a co-evolving population [48]. The tail fibers of bacteriophages also play important role to make specific contacts with receptor molecules onto the host cell surface. The common laboratory strain of bacteriophage λ, λ wild type, has been derived from the strain Ur-λ, which is the original strain of λ. This new strain carries a frameshift mutation in *stf* gene relative to Ur-λ. The Ur-λ phages have thin tail fibers which are absent in λ wild type. The expanded receptor specificity and higher adsorption rate to host cells for Ur-λ suggests the importance of tail fibers in virus-host interaction [49].

LamB protein is a well-characterized example of a porin, called maltoporin. It is required for growth on limiting concentrations of maltose. In addition to λ, LamB also serves as receptor for other phages, like K10 and TP1 [50-52]. The molecular weight of trimeric LamB is 135.6 kDa and formed by 3 identical subunits having molecular weight of 45.9 kDa [53-55]. Each subunit is formed by an 18-stranded antiparallel β-barrel, which form a wide channel (barrel) of about 2.5 nm diameter. Six loops are found at the end of barrel. Three loops, L1, L3 and L6 interact with the Loop 2 from an adjacent subunit and packed against the inner wall of the barrel and line the channel. The other loops form a compact structure at the cell surface. Six aromatic residues (Y6, Y41, W74, W358, W420 and F227) line up the channel interior and participate in carbohydrate transport [56]. Tyrosine 118 (Y118) also has a major impact on ion and carbohydrate transport through LamB. Akin to the LamB receptor, a number of studies had been carried out to understand how λ utilizes the LamB receptor to inject its DNA inside the host cell [44, 45]. Phage λ tail contains a hollow tube that consists of 32-stacked disks; each one is formed by six subunits of the major tail protein gpV, and has a 3 nm central hole. Phage λ uses this channel to eject its DNA. Protein J, synthesized from *gpJ*, directly interacts with the outer membrane protein LamB during the attachment of the bacteriophage to the surface of the cell. When the J gene from phage λ was substituted with the tail fiber gene of another closely related bacteriophage 434, the new phage binds to a different receptor OmpC, which is the receptor for phage 434 [57]. Mutations in λ phages which have compensatory effect with the LamB mutants that tightly blocks the phage λ adsorption are located in the C-terminal portions of J. A chimeric protein comprising the last 249 amino acids of J in fusion to the C-terminal end of the carrier maltose binding protein (MBP) could bind to LamB trimers, inhibited recognition by anti-LamB antibody and prevent λ adsorption. When this chimeric protein was reconstituted with either LamB or the loop deletion mutant LamB, both of which showed similar blocking of ion current in the lipid bilayer experiment [34]. In addition to these biochemical and genetic studies, the imaging studies significantly improve our understanding of phage-receptor interaction. A recent study of the spatiotemporal dynamics by which a phage λ, initially diffusing in bulk, arrives at a specific LamB site on the cell surface, provided a quantitative understanding of the initial viral-host interaction process [47].

The λ Phage LamB Target Finding Process

Target finding in biological processes is a long pursuit for both the biologist and physicist. The classical question of diffusion limited target finding on a cell surface was first successfully addressed by Adam and Delbrück [58]. They introduced the concept of reaction rate enhancement by dimensional reduction (RREDR). To explain the enhancement of adsorption rates they proposed a two-stage capture model; where free phages diffuse in three dimensions (3D), but once in contact with the cell surface this 3D motion is replaced by a two-dimensional (2D) 'random walk' on the cell surface, until captured by the receptor. This 3D+2D searching strategy initially explained the accelerated rate of the process of target finding. However to make this 2D diffusion to be advantageous, the adsorption energy needs to be strong enough to keep the phage on the surface of the bacterial cell while weak enough to allow the 2D diffusion [59]. Berg and Purcell were the first to predict the adsorption rate k as a function of receptors N on a bacterium and the calculated maximum adsorption rate (k_{max}), although, the calculated k_{max} value was significantly less than experimental findings. Further modification was done by considering the possible contribution of bacterial swimming to the search process. Better understanding of the adsorption process comes from a recent work that showed upon incubation of phage λ with *E. coli* strain Ymel (this strain carries wild type λ receptor) in a solution of 10 mM MgSO4 (pH = 7.4), the decrease of population of free phage in bulk obeys a double-exponential function with a fast and a slow decay time [60]. Both the fast and slow processes are specific to interactions between phage λ and its receptor and henceforth the interaction is an on-and-off process followed by an irreversible binding. The reversible and the irreversible binding rates are nearly independent of temperature, hence entropic in nature. However, this model is over simplified as it is based on uniform distribution of receptor sites in the cell surface of the bacteria, while studies using fluorescently labeled λ tail showed that spatial distribution of LamB in the outer membrane is not uniform. LamB accumulation is irregular and spiral in patterns, which are dynamic and depend on cell length [61, 62].

On the other hand, studies with fluorescently labeled λ phages showed a preferential binding of phages to the bacterial poles rather than covering the cell surface uniformly [63]. Such preferences of phage binding to bacteria for polar sites were also evident with other *E. coli* phages like P1, virulent coliphages T4 and T7 or with other bacteria than *E. coli* K-12, like *Yersinia pseudotuberculosis* (and with T7-like Yersinia phage A1122) and *Vibrio Cholera* (with T4 like vibrio phage KVP40) [21, 64-67]. A quantitative depiction of the search process was recently obtained from single-particle tracking experiments using fluorescently labeled phages [47].

In this study live *E. coli* cells were attached to the surface of a microfluidic chamber followed by addition of fluorescent viruses and the motion of individual phages was monitored for their free diffusion, motion on the surface of the host cells, and attachment. Agreement with the classical pictures, the phages showed reduction in dimensionality from 3D in bulk to 2D on the surface of the cell until they either irreversibly bind to a receptor, or fall off and continue their free 3D motion. Phages also exhibited an anisotropic motion pattern, with a distinctly preference to move along the short axis of the cell. The bound phages also showed a distinct preference for the poles of the cells, which can be explained by considering the ordered pattern of phage receptors, LamB, on the surface of the cell. The resemblance of organization of the LamB receptors with the features exhibited by viruses moving on the cell surface demonstrating a complex virus-host interaction resulting in an increased viral residence in receptor rich regions.

Conclusion

The adsorption of bacteriophage to a receptor on cell surface is the first step of virus-host interaction, which defines phage-host specificity and ultimately success of infection. To develop successful phage therapy, one has to understand this process in detail. Recent finding of the dependence of phage adsorption on spatial distribution of the receptor molecules onto the host cell surface may help the phage therapist to engineer phage molecule of better therapeutic output.

References

[1] Nikaido, H., Multidrug resistance in bacteria. *Annual review of biochemistry* 2009, 78, 119-146.

[2] Ho, K., Bacteriophage therapy for bacterial infections. Rekindling a memory from the pre-antibiotics era. *Perspectives in biology and medicine* 2001, 44, 1-16.

[3] Sulakvelidze, A.; Alavidze, Z.; Morris, J. G., Jr., Bacteriophage therapy. *Antimicrobial agents and chemotherapy* 2001, 45, 649-659.

[4] Marsh, M.; Helenius, A., Virus entry: Open sesame. *Cell* 2006, 124, 729-740.

[5] Mudhakir, D.; Harashima, H., Learning from the viral journey: How to enter cells and how to overcome intracellular barriers to reach the nucleus. *The AAPS journal* 2009, 11, 65-77.

[6] Smith, A. E.; Helenius, A., How viruses enter animal cells. *Science* 2004, 304, 237-242.

[7] Nikaido, H., Molecular basis of bacterial outer membrane permeability revisited. *Microbiology and molecular biology reviews: MMBR* 2003, 67, 593-656.

[8] Nikaido, H.; Vaara, M., Molecular basis of bacterial outer membrane permeability. *Microbiological reviews* 1985, 49, 1-32.

[9] Zhai, Y.; Saier, M. H., Jr., The beta-barrel finder (bbf) program, allowing identification of outer membrane beta-barrel proteins encoded within prokaryotic genomes. *Protein science : a publication of the Protein Society* 2002, 11, 2196-2207.

[10] Engelhardt, H.; Heinz, C.; Niederweis, M., A tetrameric porin limits the cell wall permeability of mycobacterium smegmatis. *The Journal of biological chemistry* 2002, 277, 37567-37572.

[11] Heinz, C.; Engelhardt, H.; Niederweis, M., The core of the tetrameric mycobacterial porin mspa is an extremely stable beta-sheet domain. *The Journal of biological chemistry* 2003, 278, 8678-8685.

[12] Model, K.; Prinz, T.; Ruiz, T.; Radermacher, M.; Krimmer, T.; Kuhlbrandt, W.; Pfanner, N.; Meisinger, C., Protein translocase of the outer mitochondrial membrane: Role of import receptors in the structural organization of the tom complex. *Journal of molecular biology* 2002, 316, 657-666.

[13] Schleiff, E.; Eichacker, L. A.; Eckart, K.; Becker, T.; Mirus, O.; Stahl, T.; Soll, J., Prediction of the plant beta-barrel proteome: A case study of the chloroplast outer envelope. *Protein science: a publication of the Protein Society* 2003, 12, 748-759.

[14] Parker, M. W.; Buckley, J. T.; Postma, J. P.; Tucker, A. D.; Leonard, K.; Pattus, F.; Tsernoglou, D., Structure of the aeromonas toxin proaerolysin in its water-soluble and membrane-channel states. *Nature* 1994, 367, 292-295.

[15] Petosa, C.; Collier, R. J.; Klimpel, K. R.; Leppla, S. H.; Liddington, R. C., Crystal structure of the anthrax toxin protective antigen. *Nature* 1997, 385, 833-838.

[16] Song, L.; Hobaugh, M. R.; Shustak, C.; Cheley, S.; Bayley, H.; Gouaux, J. E., Structure of staphylococcal alpha-hemolysin, a heptameric transmembrane pore. *Science* 1996, 274, 1859-1866.

[17] Skurray, R. A.; Hancock, R. E.; Reeves, P., Con--mutants: Class of mutants in escherichia coli k-12 lacking a major cell wall protein and defective in conjugation and adsorption of a bacteriophage. *Journal of bacteriology* 1974, 119, 726-735.

[18] Yu, F.; Mizushima, S., Roles of lipopolysaccharide and outer membrane protein ompc of escherichia coli k-12 in the receptor function for bacteriophage t4. *Journal of bacteriology* 1982, 151, 718-722.

[19] Hantke, K., Major outer membrane proteins of e. Coli k12 serve as receptors for the phages t2 (protein ia) and 434 (protein ib). *Molecular and general genetics: MGG* 1978, 164, 131-135.

[20] Riede, I.; Degen, M.; Henning, U., The receptor specificity of bacteriophages can be determined by a tail fiber modifying protein. *The EMBO journal* 1985, 4, 2343-2346.

[21] Hashemolhosseini, S.; Holmes, Z.; Mutschler, B.; Henning, U., Alterations of receptor specificities of coliphages of the t2 family. *Journal of molecular biology* 1994, 240, 105-110.

[22] Hashemolhosseini, S.; Montag, D.; Kramer, L.; Henning, U., Determinants of receptor specificity of coliphages of the t4 family. A chaperone alters the host range. *Journal of molecular biology* 1994, 241, 524-533.

[23] Osborn, M. J.; Rosen, S. M.; Rothfield, L.; Zeleznick, L. D.; Horecker, B. L., Lipopolysaccharide of the gram-negative cell wall. *Science* 1964, 145, 783-789.

[24] Wilkinson, S. G., Bacterial lipopolysaccharides--themes and variations. *Progress in lipid research* 1996, 35, 283-343.

[25] Weidenmaier, C.; Peschel, A., Teichoic acids and related cell-wall glycopolymers in gram-positive physiology and host interactions. *Nature reviews. Microbiology* 2008, 6, 276-287.

[26] Geller, B. L.; Ngo, H. T.; Mooney, D. T.; Su, P.; Dunn, N., Lactococcal 936-species phage attachment to surface of lactococcus lactis. *Journal of dairy science* 2005, 88, 900-907.

[27] Raisanen, L.; Schubert, K.; Jaakonsaari, T.; Alatossava, T., Characterization of lipoteichoic acids as lactobacillus delbrueckii phage receptor components. *Journal of bacteriology* 2004, 186, 5529-5532.

[28] Young, F. E., Requirement of glucosylated teichoic acid for adsorption of phage in bacillus subtilis 168. *Proceedings of the National Academy of Sciences of the United States* 1967, 58, 2377-2384.

[29] Frankel, R. W.; Joys, T. M., Adsorption specificity of bacteriophage pbs1. *Journal of bacteriology* 1966, 92, 388-389.

[30] Schade, S. Z.; Adler, J.; Ris, H., How bacteriophage chi attacks motile bacteria. *Journal of virology* 1967, 1, 599-609.

[31] Shea, T. B.; Seaman, E., Sp3: A flagellotropic bacteriophage of bacillus subtilis. *The Journal of general virology* 1984, 65 (Pt 11), 2073-2076.

[32] Friedman, D. I.; Court, D. L., Bacteriophage lambda: Alive and well and still doing its thing. *Current opinion in microbiology* 2001, 4, 201-207.

[33] Hendrix, R. W., Bacteriophage genomics. *Current opinion in microbiology* 2003, 6, 506-511.

[34] Berkane, E.; Orlik, F.; Stegmeier, J. F.; Charbit, A.; Winterhalter, M.; Benz, R., Interaction of bacteriophage lambda with its cell surface receptor: An in vitro study of binding of the viral tail protein gpj to lamb (maltoporin). *Biochemistry* 2006, 45, 2708-2720.

[35] Charbit, A.; Clement, J. M.; Hofnung, M., Further sequence analysis of the phage lambda receptor site. Possible implications for the organization of the lamb protein in escherichia coli k12. *Journal of molecular biology* 1984, 175, 395-401.

[36] Charbit, A.; Werts, C.; Michel, V.; Klebba, P. E.; Quillardet, P.; Hofnung, M., A role for residue 151 of lamb in bacteriophage lambda adsorption: Possible steric effect of amino acid substitutions. *Journal of bacteriology* 1994, 176, 3204-3209.

[37] Clement, J. M.; Lepouce, E.; Marchal, C.; Hofnung, M., Genetic study of a membrane protein: DNA sequence alterations due to 17 lamb point mutations affecting adsorption of phage lambda. *The EMBO journal* 1983, 2, 77-80.

[38] Hofnung, M.; Hatfield, D.; Schwartz, M., Malb region in escherichia coli k-12: Characterization of new mutations. *Journal of bacteriology* 1974, 117, 40-47.

[39] Hofnung, M.; Jezierska, A.; Braun-Breton, C., Lamb mutations in e. Coli k12: Growth of lambda host range mutants and effect of nonsense suppressors. *Molecular and general genetics: MGG* 1976, 145, 207-213.

[40] Randall-Hazelbauer, L.; Schwartz, M., Isolation of the bacteriophage lambda receptor from escherichia coli. *Journal of bacteriology* 1973, 116, 1436-1446.

[41] Scandella, D.; Arber, W., An escherichia coli mutant which inhibits the injection of phage lambda DNA. *Virology* 1974, 58, 504-513.

[42] Schwartz, M., Reversible interaction between coliphage lambda and its receptor protein. *Journal of molecular biology* 1975, 99, 185-201.

[43] Schwartz, M., The adsorption of coliphage lambda to its host: Effect of variations in the surface density of receptor and in phage-receptor affinity. *Journal of molecular biology* 1976, 103, 521-536.

[44] Wang, J.; Hofnung, M.; Charbit, A., The c-terminal portion of the tail fiber protein of bacteriophage lambda is responsible for binding to lamb, its receptor at the surface of escherichia coli k-12. *Journal of bacteriology* 2000, 182, 508-512.

[45] Wang, J.; Michel, V.; Hofnung, M.; Charbit, A., Cloning of the j gene of bacteriophage lambda, expression and solubilization of the j protein: First in vitro studies on the interactions between j and lamb, its cell surface receptor. *Research in microbiology* 1998, 149, 611-624.

[46] Golding, I., Decision making in living cells: Lessons from a simple system. *Annual review of biophysics* 2011, 40, 63-80.

[47] Rothenberg, E.; Sepulveda, L. A.; Skinner, S. O.; Zeng, L.; Selvin, P. R.; Golding, I., Single-virus tracking reveals a spatial receptor-dependent search mechanism. *Biophysical journal* 2011, 100, 2875-2882.

[48] Meyer, J. R.; Dobias, D. T.; Weitz, J. S.; Barrick, J. E.; Quick, R. T.; Lenski, R. E., Repeatability and contingency in the evolution of a key innovation in phage lambda. *Science* 2012, 335, 428-432.

[49] Hendrix, R. W.; Duda, R. L., Bacteriophage lambda papa: Not the mother of all lambda phages. *Science* 1992, 258, 1145-1148.

[50] Hancock, R. E.; Reeves, P., Lipopolysaccharide-deficient, bacteriophage-resistant mutants of escherichia coli k-12. *Journal of bacteriology* 1976, 127, 98-108.

[51] Roa, M., Interaction of bacteriophage k10 with its receptor, the lamb protein of escherichia coli. *Journal of bacteriology* 1979, 140, 680-686.

[52] Wandersman, C.; Schwartz, M., Protein ia and the lamb protein can replace each other in the constitution of an active receptor for the same coliphage. *Proceedings of the National Academy of Sciences of the United States* 1978, 75, 5636-5639.

[53] Ishii, J. N.; Okajima, Y.; Nakae, T., Characterization of lamb protein from the outer membrane of escherichia coli that forms diffusion pores selective for maltose-maltodextrins. *FEBS letters* 1981, 134, 217-220.

[54] Nakae, T.; Ishii, J. N., Molecular weights and subunit structure of lamb proteins. *Annales de microbiologie* 1982, 133A, 21-25.

[55] Neuhaus, J. M., The receptor protein of phage lambda: Purification, characterization and preliminary electrical studies in planar lipid bilayers. *Annales de microbiologie* 1982, 133A, 27-32.

[56] Schirmer, T.; Keller, T. A.; Wang, Y. F.; Rosenbusch, J. P., Structural basis for sugar translocation through maltoporin channels at 3.1 a resolution. *Science* 1995, 267, 512-514.

[57] Fuerst, C. R.; Bingham, H., Genetic and physiological characterization of the j gene of bacteriophage lambda. *Virology* 1978, 87, 437-458.

[58] Adam, G. D., M. , *Reduction of dimensionality in biological diffusion processes. In structural chemistry and molecular biology*. W.H. Freeman and Company: San Francisco, NC, US, 1968.

[59] Berg, H. C.; Purcell, E. M., Physics of chemoreception. *Biophysical journal* 1977, 20, 193-219.

[60] Moldovan, R.; Chapman-McQuiston, E.; Wu, X. L., On kinetics of phage adsorption. *Biophysical journal* 2007, 93, 303-315.

[61] Edgar, R.; Rokney, A.; Feeney, M.; Semsey, S.; Kessel, M.; Goldberg, M. B.; Adhya, S.; Oppenheim, A. B., Bacteriophage infection is targeted to cellular poles. *Molecular microbiology* 2008, 68, 1107-1116.

[62] Gibbs, K. A.; Isaac, D. D.; Xu, J.; Hendrix, R. W.; Silhavy, T. J.; Theriot, J. A., Complex spatial distribution and dynamics of an abundant escherichia coli outer membrane protein, lamb. *Molecular microbiology* 2004, 53, 1771-1783.

[63] Oddershede, L.; Dreyer, J. K.; Grego, S.; Brown, S.; Berg-Sorensen, K., The motion of a single molecule, the lambda-receptor, in the bacterial outer membrane. *Biophysical journal* 2002, 83, 3152-3161.

[64] Garcia, E.; Elliott, J. M.; Ramanculov, E.; Chain, P. S.; Chu, M. C.; Molineux, I. J., The genome sequence of yersinia pestis bacteriophage phia1122 reveals an intimate history with the coliphage t3 and t7 genomes. *Journal of bacteriology* 2003, 185, 5248-5262.

[65] Kaiser, D.; Dworkin, M., Gene transfer to myxobacterium by escherichia coli phage p1. *Science* 1975, 187, 653-654.

[66] Miller, E. S.; Heidelberg, J. F.; Eisen, J. A.; Nelson, W. C.; Durkin, A. S.; Ciecko, A.; Feldblyum, T. V.; White, O.; Paulsen, I. T.; Nierman, W. C., et al., Complete genome sequence of the broad-host-range vibriophage kvp40: Comparative genomics of a t4-related bacteriophage. *Journal of bacteriology* 2003, 185, 5220-5233.

[67] Steven, A. C.; Trus, B. L.; Maizel, J. V.; Unser, M.; Parry, D. A.; Wall, J. S.; Hainfeld, J. F.; Studier, F. W., Molecular substructure of a viral receptor-recognition protein. The gp17 tail-fiber of bacteriophage t7. *Journal of molecular biology* 1988, 200, 351-365.

Index

A

B

C

D

G

H

I

J

K

L

O

P

Q

R

S

T

U

V

W

Y

Z